TEN DAYS
IN PHYSICS
THAT
SHOOK THE
WORLD

How Physicists Transformed
Everyday Life

BRIAN CLEGG

ICON

This edition published in the UK and USA in 2022 by Icon Books Ltd
Omnibus Business Centre, 39–41 North Road, London N7 9DP
email: info@iconbooks.com • www.iconbooks.com

First published in the UK and USA in 2021 by Icon Books Ltd

Sold in the UK, Europe and Asia
by Faber & Faber Ltd, Bloomsbury House, 74–77 Great Russell Street,
London WC1B 3DA or their agents

Distributed in the UK, Europe and Asia
by Grantham Book Services, Trent Road, Grantham NG31 7XQ

Distributed in the USA
by Publishers Group West, 1700 Fourth Street, Berkeley, CA 94710

Distributed in Canada
by Publishers Group Canada, 76 Stafford Street, Unit 300, Toronto,
Ontario M6J 2S1

Distributed in Australia and New Zealand
by Allen & Unwin Pty Ltd, PO Box 8500, 83 Alexander Street,
Crows Nest, NSW 2065

Distributed in South Africa
by Jonathan Ball, Office B4, The District, 41 Sir Lowry Road,
Woodstock 7925

Distributed in India
by Penguin Books India, 7th Floor, Infinity Tower – C, DLF Cyber City,
Gurgaon 122002, Haryana

ISBN: 978-178578-834-5

Typeset in EB Garamond by Marie Doherty

Printed and bound in Great Britain by Clays Ltd, Elcograf S.p.A.

Praise for *Ten Days in Physics That Shook the World*

'Those in search of a well-written account of the world of science should look no further ... *Ten Days in Physics That Shook the World* succeeds where much of science writing fails, by creating a clear path between pivotal moments in scientific history and the world as we know it today.'
Reaction

'[A]n engaging exploration, ending with interesting speculation on the nature of a future 11th day.'
Times Higher Education

'[I]nformative and accessible ... a really good potential entry point for younger readers looking to have a big picture view on what physics is.'
Irish Tech News

'[N]icely written and highly illuminating ... an engaging read not just for engineers and scientists, but for a broad "general" audience.'
Engineering & Technology

'Clegg is a skilled wordsmith and this is a light, easy read, filled with intriguing details.'
Peet Morris, popularscience.co.uk

'[A] solid primer ... Those new to the field will find this a fine overview of touchstone moments.'
Publishers Weekly

ABOUT THE AUTHOR

Brian Clegg is the author of many books, including most recently *Quantum Computing* (2021) and *What Do You Think You Are?* (2020). His *Dice World* and *A Brief History of Infinity* were both longlisted for the Royal Society Prize for Science Books. Brian has written for numerous publications including *The Times*, *Nature*, the *Observer*, the *Wall Street Journal*, *BBC Science Focus* and *Physics World*. He is the editor of popularscience.co.uk and blogs at brianclegg.blogspot.com.

www.brianclegg.net

For Gillian, Chelsea and Rebecca

Contents

CONTENTS

Introduction

Physics is central to our understanding of how the world works. But more than that, key breakthroughs in physics – and physics-based engineering – have transformed the world we live in. In this book, we will journey back to ten key days in history to understand how a particular breakthrough was achieved, meet the individuals responsible and see how that breakthrough has influenced our lives.

It is fashionable for historians of science to criticise the idea that individuals deserve to be considered geniuses who have made a unique contribution. And, as will be made clear, it is certainly true that all of the people picked out on our ten days built their contributions on the work of others. Yet there is no doubt that until the three most recent of our days, each individual was responsible for a change that contributed to making the modern world possible.

In 21st-century physics, significant breakthroughs are often the work of big teams. The research undertaken at the CERN particle laboratory or on the LIGO gravitational wave experiment in the US can involve hundreds or even thousands of contributors. Yet historically there have been individuals whose contributions have been more than that of a standard cog in a wheel. These are people who stand out beyond their peers, however much their work may have depended on the wider canvas of thinkers of their day. And even now, although many scientists may work on a particular theory or experiment, there is often a key moment when a handful of individuals have been pivotal in making a discovery happen.

The earlier days in our journey through the history of physics involve the development of a fundamental understanding of the underlying science, while the later ones highlight physics-based engineering, involving the invention of new ways to use physics knowledge. It's not that we haven't seen new developments in pure physics that have changed our understanding of the universe since the 1950s, but most of the more recent such advances have had fewer direct impacts on our lives. Black holes or the Higgs boson, for example, are fascinating, but lack practical applications. In this book, we stay with physics and its applications that made the modern world.

We will begin back in 1687 with the publication of Isaac Newton's remarkable book *Philosophiae Naturalis Principia Mathematica*. Now virtually unreadable without professional guidance, as much for its style as for being in Latin, the publication of the *Principia* nonetheless represented a major move forward in the power of natural science. At the time, under a different type of curriculum to the present day, what Newton did was regarded as mathematics rather than physics. Even so, this was a pivotal moment in the history of science.

For our second day, we travel forward to 1831, shortly before the Victorian period, for Michael Faraday's paper on electrical induction. Recently, some of those attempting to inflate the importance of artificial intelligence have claimed that AI is 'more important than electricity'. Leaving aside the absurdity that it is impossible to have artificial intelligence without electricity, this overlooks the reality that electricity is absolutely central to modern life – and becoming even more so as we move away from fossil fuels to electricity to power, for example, cars, heating and industrial energy. Faraday's work kickstarted the practical use of what had been up to that point an entertaining novelty without worthwhile application.

We wholeheartedly enter the Victorian era on Day 3 in 1850, moving forward a couple of decades from Faraday to meet the less familiar Rudolf Clausius and explore his contribution to thermodynamics. It was thermodynamics that took the industrial revolution to a whole new level, with a better understanding of the steam engine. Equally, thermodynamics would make possible other heat engines, from the internal combustion engine to the turbines in power stations, and now underpins the mechanisms of modern heating, fridges and air conditioning. We may be losing our dependence on internal combustion, but the others remain important and thermodynamics is, at its most basic, the driving force behind life itself.

Just over ten years later, in 1861, Day 4 introduces us to Scottish physicist James Clerk Maxwell. Where Faraday gave us the means to make use of electricity, Maxwell's work opened up an understanding of the electromagnetic spectrum that includes visible light, but gives us far more. Not only did this lead to radio, microwaves, TVs and X-rays, Maxwell's legacy is typified by that wildly successful piece of technology, the mobile phone, with over 3 billion deployed worldwide.

Day 5 takes us to the final years of the 19th century and the work of the phenomenon that was Marie Curie. A woman who thrived in what was then firmly a man's world, Curie achieved remarkable things in the study of radioactivity, and the use of X-rays, with major medical benefits. On this key day, Curie revealed her most important discovery in the science of radioactive materials, radium, and set the direction for the study of radioactivity that seemed able to produce energy from nowhere.

The explanation for the source of radioactive energy came on Day 6 in 1905, with the publication of the last of a series of papers in that year that took Albert Einstein from being an obscure patent clerk to a name that would be celebrated across the world. In just three pages, Einstein showed how the special theory of relativity (published just a few months earlier) forged an unbreakable link between mass and energy, leading to the most famous equation of all time, $E=mc^2$.

For Day 7, we discover an unfamiliar name to many in the Dutch physicist Heike Kamerlingh Onnes. Working in the early years of the 20th century, Kamerlingh Onnes was the master of ultra-low temperatures and discovered superconductivity, where electrical resistance disappears, making it possible to produce the super-strong magnets required for levitating trains, MRI scanners and specialist applications such as particle accelerators.

Superconductivity is a quantum effect, and Day 8 finds us in 1947, with the viewpoint shifting from basic physics to applications of quantum theory – in particular the then rapidly developing field of electronics. It was on this day that John Bardeen and Walter Brattain, based at Bell Labs, made the earliest working transistor, the first generation of a device that would transform all our lives.

Quantum effects were also behind our Day 9 invention in 1962 of the light emitting diode (LED). This is a particularly hard event to pin down in history, as there were so many stages in the development of this technology, which is why it has only been in the 21st century, some 50 years after that date, that LEDs have become the dominant method of lighting our homes, streets and workplaces. The many subtle variants in early attempts make the choice of James R. Biard and Gary Pittman's breakthrough one of several possible key days; however, the pair have one of the best claims after producing the first commercially viable LED.

The last of our historical days in physics, Day 10, sees the first link made in 1969 in the computer network that became the internet. As with LEDs, Steve Crocker and Vint Cerf were not the only ones involved in this project, but they played a crucial role, and are the most recognisable faces of the internet's birth. It's appropriate that this breakthrough took place the same year as the first human landing on the Moon, with Neil Armstrong's famous 'One small step for [a] man, one giant leap for mankind'. What was indeed a small step forward in connecting two large computers to enable remote access became what is arguably the giant leap of the definitive technology of the modern age.

Where do we go from here? In the final chapter, we look at a handful of different possibilities for a future Day 11. Whether it will be one of these or something completely different, we can say with some confidence that there is still plenty of opportunity for physics and physics-based technologies to once again enable a world-changing innovation. For now, though, it's time to take a step back in time to a slower-paced age and Tuesday, 5 July 1687.

Tuesday, 5 July 1687

Isaac Newton – Publication of the Principia

When Newton's crowning glory, the *Philosophiae Naturalis Principia Mathematica* (Mathematical Principles of Natural Philosophy) – generally known as the *Principia* to make it less of a mouthful – was published, physics in the modern sense, making use of mathematics, was born. Featuring Newton's three laws of motion and his law of gravitation, and developed using his new and essential mathematical tool of calculus, the *Principia* set in place the mechanical principles linking forces to movement that would enable the industrial revolution to flourish, established the laws that underpin the working of jet engines and aircraft wings, and supplied the gravitational calculations needed to give us the satellites that provide everything from weather forecasts to GPS.

What makes the publication of a book world-changing? It might be at the core of a world religion or a political movement. It could be read by many millions of people and change their lives. It could be responsible for a fundamental change to society. But Isaac Newton's masterpiece fulfils none of these functions. Instead, it changed the understanding of the universe and how it works for a relatively small audience, who then spread the benefits of that understanding to the rest of us.

Interestingly, the *Principia* has something in common with novels that regularly make the 'greatest book of all time' lists – books like Proust's *À la Recherche du Temps Perdu* and James Joyce's *Ulysses*. Like them, this is a book that is widely acknowledged as being brilliant, but that very few people in recent years have managed to read (even among those of us who have attempted to do so). Yet, without doubt, this wordy Latin tome, loaded with obscure geometry, has had far greater impact than any literary masterpiece.

The year 1687

A notably uneventful year. Beyond the handful of localised European wars typical of the period, 1687 is unusual in being totally dominated by a single scientific event: the publication of Newton's *Principia*.

Newton in a nutshell

Physicist, mathematician, alchemist, heretical religious scholar, MP and bureaucrat

Legacy: Newtonian reflecting telescope, colour theory of light, laws of motion, law of universal gravitation, calculus (method of fluxions)

Isaac Newton

Born 25 December 1642 at Woolsthorpe Manor, Lincolnshire

Educated: Trinity College, Cambridge

Fellow of Trinity College, Cambridge, 1667–1696

Elected Fellow of the Royal Society, 1672

MP for Cambridge, 1689–1690 and 1701–1702

Warden of the Royal Mint, 1696–1699

Master of the Royal Mint, 1699–1725

President of the Royal Society, 1703–1727

Knighted by Queen Anne, 1705

Died 20 March 1727 at home in Kensington, London, England, aged 84

....

A new view of the universe

This is the story of a book that had a three-year genesis. Yet the history of the components that came together to make that book happen stretches back around 2,000 years. The first essential that led up to the production of the *Principia* was the developing understanding of the nature of moving objects and gravity. The second was the involvement of one remarkable man.

The physics of motion and gravitation that mostly held from ancient times through to the 17th century was built on two reasonably logical, yet incorrect beliefs. One was that an object had to be pushed if it was to be kept moving. This was apparent by observing most everyday objects. As soon as you stop pushing a cart it starts to slow down and soon it will come to a halt. An arrow that travels two hundred yards before hitting its target will arrive with noticeably less impact than one fired at point-blank range. There was, of course, one example of movement that did seem to carry on indefinitely – the motion of the planets and stars in the sky. But even this was thought to require pushing, usually as a result of divine intervention.

As far as gravity went, the accepted theory was tied into an impressively holistic picture of the universe that took in the nature of the physical elements. There were thought to be four elements that made up everything that existed below the orbit of the Moon: earth, water, air and fire. Two of these (earth and water) had a natural tendency to head for the centre of the universe, the other two had a tendency to move away from the centre. This was not a case of having a force applied to them – a natural tendency was more like a dog's natural tendency to dislike cats, an inherent part of its nature.

The tendency of earth and water towards the universal centre was described as gravity and the tendency of air and fire away from the centre was known as levity. This, incidentally, was a major underpinning of the Earth-centred view of the universe. The resistance to accepting the Sun-centred view is often portrayed as nothing more than religious obstinacy, but in fact having the Earth at the centre of everything underpinned the physics of the time, which predated Christianity and Islam. And it is anything but obvious that the heavens move because the Earth is rotating – it's easy to be critical in hindsight, but we still talk of the Sun rising and setting as if the Earth were fixed in place.

The Ancient Greek physics that lay behind this thinking had started to be questioned in medieval times by both Arabic scholars and some of the European universities, though others remained staunch in their support of the familiar old models. However, by the mid-16th century, Copernicus had argued strongly for a Sun-centred universe. This simplified the old model, which had required the fiddly invocation of epicycles, spheres within spheres, to explain the odd movement of the planets in the sky as some reverse their apparent motion because of the interaction between their orbit and that of the Earth.

This approach had been famously supported by Galileo, whose unsubtle presentation of the Copernican view in a book that appeared to mock the Pope led to his trial. However, it should be emphasised that the Copernican model wasn't the only game in town. Adopting this system implied the need to rewrite all of physics. But in the late 16th century, the great Danish astronomer Tycho Brahe had proposed a system that did away with the problematic epicycles, but still kept the Earth at the centre of things.

In what's known as the Tychonic model, the Sun, Moon and stars rotate around the Earth, but the other planets are centred on the Sun. In effect this is an accurate model of what really happens, given our viewpoint on the surface of the Earth. In the end it's a matter of where we look from (frame of reference, as physicists call it) – and at the time, the only place we had was the surface of the Earth. If you take that viewpoint, Brahe was right. It all works, but does not require a change of the fundamentals of physics.

However, Galileo did more than argue for a Sun-centred universe. It's odd in a way that we remember him for this, the work of Copernicus, and for inventing the telescope – which he didn't do as there were several earlier telescope makers. Galileo's great contribution was in reality a book he wrote after his trial while on house arrest. In *Discorsi e Dimostrazioni Matematiche Intorno a Due Nuove Scienze* (Discourses and Mathematical Demonstrations Relating to Two New Sciences) Galileo began to explore movement in the form of pendulums and balls rolled down slopes, performing experiments that ate away at the classical view of physics.

When Galileo rolled a ball down a slope, it accelerated. When he rolled it up a slope, it slowed down. It was reasonable to assume, with nothing else slowing it down, such as friction and air resistance, that a ball rolling on the flat would continue at the same speed. Just as the Copernican model required a move away from the old element-based view of gravity and levity, so this kind of exploration of the physics of motion undermined the classical idea that a push was required to keep things moving.

The Lincolnshire wonder

This was the scientific world – one where the old certainties were increasingly being questioned – into which Isaac Newton was born on 25 December 1642 in the Lincolnshire farmhouse known as Woolsthorpe Manor, into what could only reasonably be described as a troubled family.

Newton's confusing dates

25 December 1642–20 March 1726 old style
4 January 1643–31 March 1727 new style

Isaac Newton has some of the most troublesome birth and death dates in the history of science. Even respectable biographical dictionaries have been known to get them wrong. The year he was born, the year he died, the well-known fact that he was born on Christmas Day and the idea that he was born the same year that Galileo died are all up for dispute, depending on the calendar that you use.

The problem arises from England's late adoption of the Gregorian calendar, which did not take place until the 1750s. This means that when Newton was born, England was ten days behind the modern calendar, while by the time of his death, England was eleven days behind. To make things even more confusing, in the old calendar 25 March marked the start of the new year, distorting the date of Newton's death. This odd date was based on the religious feast of the Annunciation, and is why in the UK the tax year still runs from 6 April one year to 5 April the next (which, allowing for the calendar shift, are the historical new year dates).

This confusion leads to downright errors where events on one calendar are linked to events on the other, such as the media's tendency to note Newton's birthday on the modern Christmas Day ... but failing to note that the Christmas Day of the 17th century does not fall on 25 December in our calendar.

Newton's upbringing was difficult. His father died before he was born and his mother, Hannah, remarried a local rector when Newton was three, abandoning the boy to her parents so she could live with her new husband's family. We know that Newton suffered: in one of his notebooks, among his listed 'transgressions' were 'Threatening my father and mother Smith to burn them and the house over them' and 'Wishing death and hoping it to some'.

Although Hannah came back to Woolsthorpe after her second husband's death when Newton was eleven, the boy was soon parcelled off to school in Grantham, boarding with the family of Mr Clark, an apothecary in the town. Newton seems to have been initially disliked at school, but his practical skill at building mechanical models brought at least a degree of acceptance, even if he was never popular.

Trouble with Newton's mother would continue, as she removed him from school to work on the farm. Newton regularly sought out opportunities to escape farm work and read; eventually, no doubt frustrated, his mother was persuaded to let him return to school when the headmaster excused Newton the 40-shilling fee usually charged to boys who came from outside the town. However, she would not support him when he went up to Cambridge, requiring him to take a position as a sizar where his keep was supported by acting as a servant to other students.

Cambridge and the Royal Society

Being a student at Cambridge required a profession of the Anglican faith in this period. We are used now to many scientists being atheists, but in Newton's day, Christianity was an expected part of life in Britain and totally integrated into the thinking of European scientists. Newton was a devout Christian, but his beliefs started out with a more puritanical flavour of Christianity than was common in the Church of England, and he would develop beliefs that were considered outright heresy by the standards of the day. It was also the norm that university fellows had to be single (this, at least, was not a problem for Newton) and ordained in the church – Newton obtained special dispensation from the King to avoid the latter requirement.

Over time, Newton's beliefs strayed into Arianism. This was a doctrine originated by a 3rd-century Libyan priest named Arius, which rejected the conventional Christian concept of the Trinity and believed that Jesus was created by God, rather than existing from the beginning. Although there had been Arian churches historically, this was an unusual belief in Newton's day, which he combined with an obsession with finding arcane meaning in ancient texts, culminating in the belief that the date of the end of the world would be no earlier than 2060, obscurely deduced from prophesies in the Bible books of Daniel and Revelation.

Newton's non-conformist attitude to his religion was of a piece with the approach he took to science. At the time, the curriculum at Cambridge was primarily based on classical sources with little encouragement to question the wisdom of authority. Galileo's books, for example, were too revolutionary to be found in the university's libraries. But Newton's approach echoed the motto of the Royal Society, which would become such a big part of his life: *Nullius in verba* (take no one's word) – effectively, question everything. And there was plenty to question in a view of physics that had changed relatively little since the time of Aristotle. Newton wasn't the first to challenge scientific authority – as we have seen, Galileo and others had done so – but he took the questioning to a new level. Newton was not a person who would go with the flow. Both in his experiments and his increasing deployment of mathematics, he was prepared to go further, to stand out from the crowd.

Newton's early scientific work was primarily on light. He was elected as a fellow of the Royal Society thanks to his construction of a reflecting telescope, but was soon at odds with the Society's Curator of Experiments, Robert Hooke, who criticised Newton's theories on colour. Hooke's (mostly incorrect) negative remarks drove Newton to threaten to resign.

Of myths and personality

The battle with Hooke would develop into a lifelong feud. There seems little doubt that it was real. It seems likely, for example, that Newton was responsible for the destruction of Hooke's portrait, leaving us without a contemporary image of a great scientist in his own right. Newton's relationships with others were often prickly and, given the concerns of the day, sometimes difficult to be sure of in retrospect. This comes across most clearly in uncertainties about Newton's sexuality.

This was a time when homosexual thoughts, let alone behaviour, would have been considered deeply sinful. Yet there is no evidence that Newton had any interest in the opposite sex. The only women other than his mother with whom he had any notable connection were Catherine Storer, the stepdaughter of the apothecary in Grantham, who claimed after Newton's death that he had considered marrying her, and his half-niece Catherine Barton, who acted as his housekeeper towards the end of his life. By contrast, Newton certainly had a close relationship with John Wickins, with whom he shared accommodation for over twenty years.

Another relationship would develop with the much younger Swiss mathematician Nicolas Fatio de Duillier. For over five years the pair exchanged affectionate letters and Newton not only gave the younger man gifts but gave voice to more feelings in his letters than he ever recorded elsewhere. If there was a relationship, it seems to have ended abruptly after a visit Newton made to Fatio in London in 1693 – this may have been because Newton felt Fatio was being indiscrete about their shared enthusiasm for alchemy, and almost certainly contributed to Newton's imminent breakdown. For the next few months, Newton wrote to a number of his acquaintances telling them he wanted no more to do with them, even suggesting the philosopher John Locke had attempted to embroil Newton with women. He seems to have recovered quickly, but clearly this was a man under stress.

To modern eyes, the above-mentioned fascination with alchemy might also imply a troubled personality. Indeed, it was something of an obsession for Newton, dominating much of the period of his life when he made his achievements in physics. And it is possible that the materials he worked with, such as mercury, may have contributed to his breakdown. But the study of alchemy, though considered dubious (and if used in certain ways illegal) was not incompatible with the scientific thought of the day, and fits well with a mind that clearly straddled scientific and mystical religious thought.

A final example of the uncertainty around personal detail is in what is surely the most famous story concerning Newton: the apple. Despite some modern assertions to the contrary, Newton's apple is not entirely mythical (though any idea of it landing on his head is). The source for the story of the apple is Newton himself, quoted by his younger contemporary, William Stukeley. In his book *Memoirs of Sir Isaac Newton's Life*, Stukeley describes paying a visit to Newton in 1726 at Newton's lodgings in Orbol's Buildings in Kensington. Stukeley tells us that they were sitting under apple trees in the garden after dinner (drinking tea) and Newton claimed that he had first thought about the nature of gravitation 'occasion'd by the fall of an apple'.

Some suggest that this late revelation, when Newton was in his eighties, was an attempt to build up his own mythology at a time when Newton probably had no real recollection of the events: certainly, there is no earlier record of the apple incident. However, it is a perfectly reasonable assertion. To deny it seems more an attempt to be iconoclastic for the sake of it than a genuine concern for the truth. What certainly is true, though, is that Newton never had a strong urge to publish his work, often not making it public for many years. And this would be true of at least some of the contents of the *Principia*.

A reluctance to publish: the 1687 day

It has frequently been stated that Newton developed calculus and his theory of gravitation in a concentrated period of under two years from 1665 when dispatched home from Cambridge during an outbreak of plague. This is a wild exaggeration. He was certainly slow to publish his work on forces and gravitation, but when the *Principia* was eventually published it pulled together material he had been working on for over 20 years. Although Newton's early work was sent to the Royal Society, after the criticism from Hooke, Newton refused to send in details of further theories on light, which he held back from the 1670s all the way through to the publication of his book *Opticks* in 1704. The factors that shaped Newton's personality seem to have inclined him to secrecy. Although he was determined to be recognised as the first to come up with ideas, he resisted publication at every opportunity.

That the *Principia* was published at all was as much down to astronomer Edmund Halley as to Newton. Halley, Hooke and the polymathic architect of St Paul's Cathedral, Christopher Wren, had been talking about planetary motion in a London coffee house in 1684 when Hooke claimed to have proved that the force keeping the planets in their orbits decreased with the square of the distance between the planet and the Sun. Clearly suspecting more than a little boasting on Hooke's part, the wealthier Wren offered a reward to Hooke if he could produce this proof in two months. When Hooke failed, Halley headed up to Cambridge to speak to Newton on the matter.

Newton also claimed to have performed the appropriate calculation showing such a force would produce elliptical planetary orbits, but couldn't find where he had written it down. Three months later, he sent Halley nine pages on the topic, an undertaking that seems to have triggered the writing of the *Principia*. By the time the book was published in July 1687 it had become three volumes of Latin. The first, *De Motu Corporum* (On the Motion of Bodies) introduces basic concepts such as mass, gives us Newton's three laws of motion, and provides calculations to support the elliptical, inverse square law motion of the planets.

The second volume, arguably the least significant, called – with a snappiness of titling later rivalled by Hollywood – *De Motu Corporum Liber Secundus* (On the Motion of Bodies, Book Two), adds resisting mediums such as air and considers pendulums, waves and vortices. And the third volume, *De Mundi Systemate* (On the System of the World), features Newton's law of gravitation, describing a 'universal' force that is equally responsible for the fall of the famous apple as it is for the keeping the Moon in orbit around the Earth and the planets around the Sun.

As we have seen, and unlike his later book *Opticks*, which was written in English, the *Principia* was written in Latin. This had been the standard language of European academia in the early days

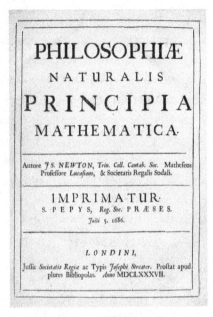

Title page of the Principia.

of universities, enabling scholars across Europe to move freely between universities and to share ideas. Academic books had been published in Latin for centuries. The use of Latin had indeed allowed an international readership, but it had also excluded the majority of the literate population from reading these books – something that was actively encouraged by many natural philosophers, who liked to quote remarks such as 'It is stupid to offer lettuces to an ass, since he is content with his thistles'. There was a deliberate attempt in some quarters to keep the revelation of arcane matters from the common herd.

But this attitude was changing. Galileo, for example, wrote his key books in Italian, not in Latin. There was a parallel with the publication of the Bible in the modern languages of the day. One of the great moves of the Reformation of the 16th and 17th-century

Christian church had been the shift from services and the Bible being in Latin, and hence not accessible to the masses, to being in the native language. Newton, however, suppressed the publication of the *Principia* in English until shortly before his death.

Sticking entirely to Latin was a change of heart. Newton had originally envisaged publishing two maths-laden Latin volumes – the first covering the material on forces and motion that ended up in volumes one and two, and the second on planetary motion – followed by a third volume in English, addressed to a more popular audience so that his work could be appreciated by the wider public. But he then deliberately made the third book less approachable, so that it could only be read by those who had mastered the principles of the first two volumes. We know this is the case as Newton starts volume three by admitting it. In part, he says, this was because 'those who have not sufficiently grasped the principles set down here will certainly not perceive the force of the conclusions, nor will they lay aside the preconceptions to which they have become accustomed over many years'.

This last, key volume was in danger of being lost entirely. In 1686, Newton's old foe Robert Hooke heard an extract read at the Royal Society and complained bitterly that Newton did not give him the credit he felt he deserved for ideas he had originated on gravitation. Newton wrote to Halley that he intended to suppress the third volume, because 'Philosophy is such an impertinently litigious Lady that a man had as good be engaged in Law suits as have to do with her'. Halley calmed Newton down: the result was that the book was finished in April 1687, though by then Newton had worked through all three manuscripts, carefully deleting almost every mention he had previously made of Hooke.

The intention was that the Royal Society would pay for the publication of the *Principia*, but infamously the Society squandered its budget on the publication of a far from memorable book by Francis Willughby called *De Historia Piscium* (On the History of Fish). As a result, the *Principia* was in danger of not appearing until Halley, who had already nursed along the production, volunteered to pay for the printing of the 400 to 500 copies of the first edition. This wasn't necessarily entirely altruistic as it has been suggested that Halley probably made a small profit from the venture. He also wrote a lavishly favourable review for the *Philosophical Transactions of the Royal Society* and wrote an opening ode to Newton for the book (something perhaps we should see more of in modern scientific works), including the fulsome lines below.

Part of the *Ode on This Splendid Ornament of Our Time and Our Nation, the Mathematico-Physical Treatise by the Eminent Isaac Newton*

Edmund Halley, translated by I. Bernard Cohen and Anne Whitman

Behold the pattern of the heavens, and the balances of the divine structure;
Behold Jove's calculation and the laws
That the creator of all things, while he was setting the beginnings of the
   world, would not violate;

...

O you who rejoice in feeding on the nectar of the gods in heaven,
Join me in singing the praises of NEWTON, who reveals all this.
Who opens the treasure chest of hidden truth,

...

No closer to the gods can any mortal rise.

Translation copyright © 2016 I. Bernard Cohen and Anne Whitman

A new physics

Much of the *Principia* is difficult to read, but to those who managed to work through the hundreds of pages, it proved transformative. As we have seen, the big steps forward it produced included the laws of motion and the law of universal gravitation. Also significant was the concept of mass.

Mass, as opposed to weight, was an idea that Newton introduced – one that is essential to properly understanding the relationship between force and motion, but that even now is often not well grasped. Mass is an inherent property of a piece of matter with two distinct functions: it fixes the inertia of a body and determines how it will respond under the attraction of gravity. Strictly speaking these functions could have distinct values, but in practice the same mass applies to both.

Inertia is the natural resistance of a body to movement. The more mass it has, the more force it takes to accelerate it. Newton's stroke of genius here was moving away from the concept of weight, which is the force applied to a mass at a particular level of gravitational attraction. Familiar experience now, let alone in Newton's day, is only of the weight an object has on the Earth's surface – but that same object will have a totally different weight in space, or on the Moon. In reality, weight should be measured in units of force (newtons in the scientific system), but in practice we use the unit of mass to stand in for it. So, when we say something weighs a kilogram, we really mean it has the weight that such a mass has on the surface of the Earth.

Newton's laws of motion also required an ability to look beyond the obvious. The ancient Greeks had reasonably assumed that something needs to be pushed to keep it moving: after all, it fits well with what we observe. But Newton's first law says that an object will stay as it is, in motion or stationary, unless a force acts on it, recognising that forces such as friction and air resistance slow down moving objects. As we have seen, Galileo had also recognised this (even Aristotle had suggested it would be the case if a vacuum could exist, in his argument against the existence of vacua), but it was Newton who solidified it as reality.

The second law of motion gives a relationship between the way that an object's motion changes, the force applied to it and the mass of the object. Although Newton didn't specify it in this way, we would now say that force is equal to the mass times the acceleration produced. Galileo had experimentally established aspects of this relationship, but once again it was Newton who pinned it down to mathematical certainty.

Finally, the third law tells us that an action has an equal and opposite reaction. When you push on something, it pushes back on you. This is pretty obvious if, for example, you push a wall, yet it was a novel observation when applied more generally, one that underlies many physical interactions, not to mention being responsible for the way that aircraft engines, aircraft wings and rockets function.

Mass and the laws of motion are the bread-and-butter stuff of the *Principia*, at the core of the many examples in the first two books. But Newton's greatest leap was in the development of the concept of universal gravitation. One essential component leading to this in the *Principia* was the 'shell theorem' which proves that the mass of a body can be considered to be acting at a point, the body's centre of gravity. Newton's law of gravitation then relates the force between two bodies to their masses and the inverse square of the distance between those centres. And it includes the realisation that the same force can be responsible for the falling apple and an orbiting body, which Newton elegantly shows by imagining bringing the Moon's orbit down to nearly touch the Earth's surface.

Although he did have a theory as to how gravitation worked, Newton explicitly said in the *Principia* that he would not explore this (or any) hypothesis ('*hypotheses non fingo*' in the original Latin). The requirement of his theory for bodies to influence each other at a distance was described at the time as occult, meaning hidden, because there was no obvious mechanism by which it could work. It would not be until Einstein's development of the general theory of relativity more than 300 years later that Newton's work would be refined to more accurately reflect reality and an explanation would be provided for gravity's action at a distance.

Underlying the *Principia* is another of Newton's great pieces of original thinking – calculus. Although the vast majority of the mathematical calculations on show within the book are made geometrically, there is no doubt that calculus, the mathematics of change that is especially suited to the kind of varying acceleration necessary to deal with the mechanics of movement and gravitation, was central to the development of his theory. Newton does include its use, but far less than he could have done. However, calculus has to take something of a back seat in our exploration of this particular day, both because it was developed in parallel by Gottfried Leibniz (devising the terminology and notation we still use today), and because its use is mostly hidden in the *Principia*.

Newton, the person

Newton is regularly placed on a pedestal with a handful of others as a transformative genius, yet science is a collaborative process in which no one operates in isolation. Newton himself appeared to highlight this in his famous quote 'If I have seen further it is by standing on ye shoulders of giants', though this is now generally thought to have been a put-down. It appears in a letter to Robert Hooke, who, as we have seen, Newton despised – it's hard not to think that the remark was not linked to Hooke's reputation of having a hunched stature, making him anything but a giant physically.

Later, Newton would have feuds with the mathematician Leibniz over priority on developing the calculus and with the Astronomer Royal John Flamsteed, who provided data to Newton to help confirm his work on gravity. The pair would have a very public falling-out over an astronomical catalogue which Newton pressured Flamsteed to produce, only to publish an unauthorised version himself before Flamsteed had finished it.

However, despite living at a time when much new scientific thinking was in the air, Newton was one of the more isolated workers in the field. During his active period as a scientist, he did not frequent the talking shops of the scientific world and during his productive period had a distinctly bumpy relationship with the organisation he would eventually head up, the Royal Society. While he certainly made use of whatever he could lay his hands on, his approach to physics was radically different to his great predecessor Galileo, with mathematics far more firmly at its heart.

We should also take into account the fact that although Newton put in some periods of effort where he concentrated on physics and mathematics, for the majority of his life it was not his primary concern. The catalogue of his personal library underlines this. He owned around 2,100 titles on his death, a very large number for the period, yet of these only 109 were on physics and astronomy and 126 on maths, compared with 477 theological titles. After the publication of the *Principia*, he spent far more time on his political life and engagement with the Royal Mint – where he brought forensic efficiency to a crackdown on those who were counterfeiting and trimming metal from coins – than on science.

It is frequently said that Newton was the first person knighted for his science work, but in reality, his knighthood was for his political activities and his work at the Mint. Considering this, the scale of his achievement underlines his rightful claim to genius – and the zenith of that achievement was in the publication of the *Principia*.

Life changers

There are many places where Newton's work has fed into later developments in physics, but these are a few examples where specific outcomes have been driven by the contents of the *Principia*.

Mechanical engineering

All mechanical engineering makes use of Newton's laws of motion. Although it was possible to produce machines before – the most basic machines, such as the wheel, for example – the development of modern machinery requires frequent use of these essentials.

Jet engines

The jet engine (and the rocket) depends entirely on Newton's third law of motion. The engine pushes air and fuel exhaust out of the back and the result is that the engine (and hence the plane it is attached to) is pushed forward.

Wings

The lift effect of aircraft wings is often attributed to Bernoulli's principle, where the shaping of the wing changes air pressure, resulting in lift. However, the simplest way of looking at an aircraft wing's action is that it is angled in such a way that it pushes air downwards as it moves through it. The third law means that the wing is accordingly pushed upwards.

Satellites

Although the most dramatic aspect of spaceflight has been the ability to send astronauts away from the Earth, by far the biggest impact on everyday life comes from the satellites that provide us with communications, weather forecasts, GPS navigation and more. It would have been impossible to put satellites into orbit (or, for that matter, to put people to the Moon) without making use of Newton's work on gravitation.

Thursday, 24 November 1831

Michael Faraday – Reading of 'Experimental Researches in Electricity'

I n contrast with Isaac Newton, Michael Faraday was a self-effacing individual, well aware of his own limitations. However, the primarily self-taught Faraday had a natural grasp of physics and would provide others with huge insights by devising the concept of fields, an approach that would transform theoretical physics. Despite a fierce attack from his mentor, Humphry Davy, who thought Faraday had stolen a discovery from Davy's friend William Wollaston, Faraday set in motion the development of the key electrical devices that would bring electricity to the masses. This culminated in his discovery, presented in November 1831, of electrical induction. Faraday made the electrical motor and generator practical. In truth, Elon Musk should have called his car company Faraday, not Tesla.

The year 1831

This year saw the publication of *The Hunchback of Notre-Dame* by Victor Hugo, the physical location of the North Magnetic

Pole established, the coronation of King Leopold of the Belgians after Belgium's secession from the Netherlands the previous year and that of William IV of the United Kingdom, the departure of the HMS *Beagle* from Plymouth with Charles Darwin on board, and the birth of the great Scottish physicist James Clerk Maxwell who would become Faraday's scientific successor on the subject of electromagnetism.

Faraday in a nutshell

Physicist, chemist and science communicator

Legacy: use of fields in physics, electromagnetism, electrical engineering, electrical induction, electrolysis, discovery of benzene

Michael Faraday

Born 22 September 1791 in Newington Butts, London, England

Educated: primarily self-educated

Bookbinder's apprentice, 1805/6–1812

Assistant to Humphry Davy at the Royal Institution, London, 1813

Assistant Superintendent of the House at the Royal Institution, 1821

Married Sarah Barnard, 1821

Elected to the Royal Society, 1824

Director of the Laboratory at the Royal Institution, 1825

Initiated Royal Institution Christmas Lectures, 1827

Fullerian Professor of Chemistry at the Royal Institution, 1833

Died 25 August 1867 in Hampton Court, Middlesex, England, aged 75

Magic action at a distance

For thousands of years, electricity and magnetism had been known as mysterious phenomena that could cause movement from a distance and create sparks in a seemingly magical fashion, but Michael Faraday would build on a range of experiments in the early 19th century to study the interaction of electricity and magnetism, resulting in his 1831 paper on the crucial phenomenon of electrical induction.

Despite clear similarities in their behaviour, electricity and magnetism were treated as separate concepts right up to the 19th century in physics and far later in the wider world – at school and in our everyday experiences, we still tend to act as if electricity and magnetism were unrelated, even though they are both aspects of the overarching phenomenon of electromagnetism.

Because awareness of their existence predates modern science, it would be hard to say which aspect of electromagnetism was recognised first. We know that the Ancient Greeks, for example, were aware of both static electricity and magnetism. Static electricity is the build-up of electrical charge on an object, often through triboelectricity (literally 'rubbing electricity'). We experience this when, for example, we rub a balloon, making it stick to the wall or capable of picking up small pieces of paper. Similarly, build-ups of static electricity cause a shock and a crackle when we take off a piece of clothing made of artificial fibres.

In Ancient Greek times, the go-to material for producing static electricity was the fossilised tree resin called amber. Its effect was noted by Thales of Miletus around 600 BC, giving us a familiar word: in Greek, the word for amber was *elektron*. As for magnets, with their ability to attract some metals and to orient themselves north–south, the Greeks were aware of a particular type of stone from a region of

Greece called Magnesia. They referred to this as Magnesian stone – *magnetis lithos*.

Garlic and goats' blood

It's now hard to envisage electricity and magnetism as anything other than scientific phenomena, but the Ancient Greeks had stronger links to a magical viewpoint than a scientific one. Nothing makes this clearer than the relationship of magnets with garlic and goats' blood. Both the Ancient Greeks and the Romans believed that rubbing a piece of garlic on a magnet would stop it from working, after which it could only be reactivated by dipping it in the blood of a goat.

The reason we now struggle with understanding this assertion is the powerful link we have in our minds between experiment and our understanding of nature. It seems painfully obvious that it would be easy to rub a magnet with garlic and establish that the magnet still works. (It wouldn't surprise me if one or two readers of this book give it a try – I can confirm that my fridge magnets are entirely unsusceptible to garlic treatment. As a result of this, I thankfully have no need to test the goat's blood hypothesis.)

The two keys to understanding this apparently bizarre belief were the dependence on philosophy as a means of understanding the world around us, and the extreme respect that was awarded to authority figures. Often, accepted wisdom would be decided by philosophical debate, after which the winning argument was taken as fact until eventually a rebel would take the extreme step of testing it out.

The importance of experiment and experience over received wisdom was championed in the Arabic-speaking world towards the end of the first millennium and was taken up by some European thinkers. For example, the 13th-century English friar and natural philosopher Roger Bacon stressed the essential nature of experiment. Although Bacon himself mostly only experimented with light, he was inspired in championing experiment by the French author of *Epistola de Magnete* (Letter on Magnetism), Peter of Maricourt, or Peter Peregrinus. Bacon appears to have met Peter while at university in Paris and admiringly wrote of him: 'He gains knowledge of matters of nature, medicine and alchemy through experiment ...'

Accordingly, a whole section of Bacon's masterpiece the *Opus Majus* (Great Work), a vast proposal for an encyclopaedia of knowledge, was dedicated to the importance of experiment as the means to test a philosophical theory. Bacon wrote: 'He, therefore, who wishes to rejoice without doubt in regard to the truths underlying phenomena must know how to devote himself to experiment. For authors write many statements, and people believe them through reasoning which they formulate without experience. Their reasoning is wholly false.'

Admittedly, Bacon's idea of what constituted an experiment was probably closer to what we would now call experience – but the fact remains that he was unusual in arguing for the need to test things out, rather than rely on the power of philosophical argument alone. So powerful was the old way that, as we saw in Day 1, even in Isaac Newton's youth, much 'science' still related back to the views of the Ancient Greek philosophers. So, despite our incredulity now, the belief of the effects of garlic and goats' blood on the power of magnets was still common in the 17th century.

The suggestion was not without its doubters, though. The Italian author of *Magiae Naturalis* (Natural Magic), Giambattista della Porta, wrote (from a 1658 translation of his 1589 book): 'Not onely breathing and belching on the Loadstone* after eating of Garlick, did not stop its vertues; but when it was all anointed over with the juice of Garlick, it did perform its office as well as if it had never been touched with it.'

The idea that garlic would work against a magnet derives from the doctrine of sympathy and antipathy – that some things in nature have a natural sympathy or antipathy towards each other. The reason garlic was thought to be bad for magnets was exactly the same reason it is associated in folklore with warding off vampires. Garlic was considered antipathetic to poison and the power of a magnet was thought in some way to be poisonous. Similarly, goats' blood was thought to be sympathetic to a magnet.

Della Porta also provided a helpful guide to some of the uses of magnets, noting, for example, that putting a magnet with an image of Venus engraved on it under your wife's pillow would act as a test of faithfulness: if she was faithful, she would be attracted to you in her sleep; if not, she would push you out of bed. Whether or not he subjected this hypothesis to experimental verification is not clear, but it appears that at least some of his more accurate work on the science of magnetism was stolen from another researcher.

Della Porta seems to have lifted some of his material from fellow Italian Leonardo Garzoni. In a treatise, Garzoni described a range of experiments with magnets and iron bars, experiments that would also find their way into a book published in 1600 called *De Magnete* (On the Magnet) by the English natural philosopher William Gilbert,

---

* Magnets were referred to as loadstones or lodestones. 'Lode' was an Old English word for a road or way, so a lodestone pointed the way when made into a compass needle.

which would not only provide the groundwork for our scientific understanding of the magnet's action (if not how it works), but arguably – just as Peter Peregrinus' work had inspired Roger Bacon – shaped the development of experimental science itself. In fact, it was Gilbert's book that seems to have inspired Galileo to indulge in scientific experiments.

It was Gilbert who clearly identified that compasses act in the way they do because the Earth itself is a magnet, making little magnetic balls he called terrellae to experimentally compare with the interaction between compasses and the Earth.

Electrical experiences

Electricity was even more familiar than magnetism in some of its manifestations – lightning, for example – but initially there was no association between, say, the static electricity generated by rubbing amber and the vast electrical discharges in the sky. There clearly were similarities between the effects of static electricity and magnetism, but they were also quite different – magnetism, for example, only attracted iron, where static electricity attracted a range of substances from paper to hair. As well as experimenting on magnets, Gilbert carried out a number of experiments on electricity, extending the range of materials that could generate electrical effects (though noting that metals did not produce electricity, clearly separating magnets and electrical objects).

During the 18th century a range of triboelectric devices were developed that produced a more sizeable electrical charge than rubbing amber, and demonstrations of electrical phenomena – culminating in the 'electrical boy' where a youth suspended from insulating materials was used to conduct electrical effects – became popular entertainment. One significant step forward of the period was Benjamin Franklin's famous kite experiment that linked lightning to earthbound electricity.

Franklin's kite

It might seem obvious to us, with our experience of all things electrical, that lightning has to be some form of electricity, but it is only relatively recently that this was realised. Famously, the American politician and scientist Benjamin Franklin is said to have experimented with the nature of lightning by flying a kite in a thunderstorm in 1752. The kite is supposed to have picked up the electrical charge from the storm which caused a spark to jump from a key Franklin fixed to the kite string. However, if true, this was a very risky operation that could have ended in death – kites should never be flown in a thunderstorm.

This experiment has a murky history. We don't know for certain that Franklin ever performed it. He certainly proposed trying something like this in a publication issued in 1750, and others did undertake it, but there is no contemporary documentation of Franklin himself doing so. If he did, it's unlikely that he flew a kite and waited for it to be struck by lightning, as the experiment is often portrayed. Instead, his proposal was to tap into the electrical charge in the thunderclouds to cause a build-up of electricity on a key, with no lightning strike taking place. The charge was then to be passed using a wire to a primitive storage device called a Leiden jar, where it could be demonstrated that the power of the storm behaved exactly like ordinary electricity that was generated on the ground.

Up to this point, electrical considerations were primarily of static electricity – a build-up of electrical charge, whether in a cloud or on a piece of amber, which might then produce a brief flow of electrical current in the form of a spark. But the first step that was necessary to enable Faraday's work was the ability to produce a consistent flow of current electricity – what we now know to be a flow of electrons through a conducting material. This became possible with the construction of an electrical battery by the Italian scientist Alessandro Volta. Early batteries were called electrical piles because they literally consisted of a pile of cells, each cell being a copper disc, then a disc of paper soaked in salt water, then a zinc disc – a combination of materials that undergo a chemical reaction to produce a flow of electrons.

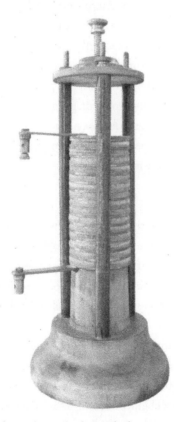

An electrical pile.

The final steps leading to 24 November 1831 were a series of experiments by European scientists that began the realisation of the relationship between electricity and magnetism – not in the form of the similarity noted by early observers, but rather in the way that one of these phenomena could influence the other. The Danish scientist Hans Christian Ørsted had shown in 1819 that a compass needle could be moved by a nearby wire carrying an electrical current, indicating that electricity had a magnetic effect. Two years later, French scientist André-Marie Ampère extended the science he called electrodynamics to note that wires carrying electrical currents could be made to attract or repel each other, as if they were magnets. A third observation that would contribute to Faraday's work was the discovery in 1824 by French scientist and politician François Arago that a rotating copper disc would drag around a magnetised needle suspended above it. Clearly this was not a magnetic effect – copper is not a magnetic material – but something happened in the metal to produce this turning motion.

A modest man

Many Victorian scientists were wealthy, able to indulge in a passion for science because they had no obligation to earn money. None could be further from this breed of scientist than Michael Faraday. Not only did he come from a poor family, he would turn down a number of honours, remaining plain Mr Faraday until his death.

Faraday's parents had moved to London from the north of England before he was born, in a search for work. Young Michael had a limited schooling before being apprenticed at age fourteen to the French-born bookbinder George Riebau. There is no doubt that Riebau was one of two major influences on Faraday's intellectual development. A refugee from the French revolution, Riebau encouraged Faraday to read the books in the shop. This, along with a fascination with science picked up when attending talks at the City Philosophical Society, set Faraday on a road that took him under the sway of his second mentor: Humphry Davy.

Davy was himself from a relatively poor background, but had attended a grammar school in his native Cornwall. After being apprenticed as an apothecary, Davy was able through a chance meeting to take a step up at the Pneumatic Institute, a medical research centre in Bristol, and then a further elevation at the Royal Institution in London, where his spectacular public lectures won him a popularity which, coupled with his marrying a wealthy widow, saw him transformed into a gentleman.

Faraday had bound the lecture notes he made at meetings of the Philosophical Society, which sufficiently impressed a client of Riebau's that the client gave Faraday tickets to a series of Davy's lectures. Faraday was even briefly able to act as an assistant to Davy when the older man's eyesight was temporarily damaged in a chemical explosion. Faraday was then returned to the bookbinders – but a longer-term position came up when the lab assistant at the Royal Institution was fired for brawling, and, with his previous experience, Faraday was a natural for the post.

The black ball

In some ways Davy was excellent for Faraday's career, taking him on a European tour where they met up with well-known scientists, though Faraday was never allowed to forget his position, required to act as valet to Davy as well as scientific assistant. Even so, Faraday progressed. In 1821, he married Sarah Barnard, a member of the same dissenting church, and they moved into a family suite of rooms at the Royal Institution. At the time Faraday would probably have been described as a solid worker – not flashy, not given to bursts of creativity, but painstaking at his tasks. However, Davy was about to turn on his former assistant.

Faraday had been working in chemistry, but he was asked by Davy to pull together a picture of the developments in the new and exciting field that would become known as electromagnetism. As we will see in his 1831 paper, Faraday was well aware of experiments that had been undertaken by his European counterparts and set out for a considerable part of 1821 to reproduce their results, so that he could better assemble his information. In the process, something unexpected happened. While the earlier research had shown a simple electromagnetic attraction, when Faraday sent electricity through a wire next to a permanent magnet, the wire moved. If that wire was suspended so it could freely rotate, it started to circle the magnet.

This was a dramatic sign that the interaction between electricity and magnetism was not just a fascinating piece of physics, but had potential for practical applications – this was, in its simplest form, an electric motor. Given that Faraday was Davy's protégé, it might be expected that the older man would celebrate Faraday's success and do everything that he could to support it. Instead, Davy attacked him.

The problem seems to have had social status at its root. A friend of Davy's, William Wollaston, had come up with an unsupported hypothesis that electrical current spiralled around a wire as it travelled along. Wollaston was convinced that Faraday's discovery was a direct outcome of his hypothesis and accused Faraday of stealing his idea. Davy supported Wollaston. Despite his own humble origins, Davy now regarded himself part of the establishment, as was Wollaston.

Faraday remained an outsider. The rift between the two was never truly healed. When in 1824 Faraday was proposed for fellowship of the Royal Society, only one individual made use of the black ball indicating a rejection: Davy.

Introducing induction

For some time after the Wollaston affair, Faraday stayed away from electromagnetism, returning to his first love of chemistry, dealing with the administration of the Institution and expanding the public lecture programme to include Friday night formal events and the Christmas lectures for young people that remain popular to this day. However, mysteries like Arago's disc were too fascinating to leave alone for ever, and Faraday returned to the field in 1831.

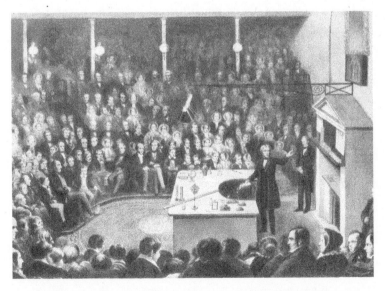

Faraday delivering a Christmas lecture at the Royal Institution.

In August of that year, he discovered that by wrapping two windings of insulated wire around the straight sides of a loop of iron shaped like a link in a chain, he could generate a current in one wire by sending a current through the other – despite there being no direct contact between the two. The fascinating thing about this so-called electrical induction was that the new, induced current did not flow constantly. It surged briefly when the current through the first wire was switched on, then went away. Similarly, it surged into brief existence when the current was switched off.

As Ampère had shown, an electrical current had a magnetic effect. So, when the current was sent through the first wire it would have a magnetic effect on the other. It seemed that having a changing level of magnetism resulted in an electrical current being induced. To test this out, Faraday experimented with moving a permanent magnet near a wire and found that this too induced a current. Just as his moving wire presaged the electric motor, this provided the foundation for the electrical generator or dynamo.

The Peel rejoinder

It is said that when Faraday demonstrated his discovery of electromagnetic induction, the then British prime minister Robert Peel asked him what use his discovery was, to which Faraday responded, 'I know not, but I wager one day your government will tax it.'

There are significant doubts about the accuracy of this story. It's sometimes said to be a comment made to the then chancellor of the exchequer, William Gladstone, rather than the prime minister, while the savvy dig at the world of politics seems quite distant from Faraday's usual lack of worldliness – and the quotation comes in a range of forms including 'Why sire, there is the probability that you will soon be able to tax it'.

Peel was prime minister twice, from 1834 to 1835 and 1841–46, while Gladstone did not become chancellor until 1852. Although the dynamo was not produced in a modern form until 1866, early dynamos were in use from the 1840s, which makes it seem unlikely that such an exchange could have occurred except in Peel's first term.

Although this development cemented the importance of Faraday's work for the world, and would soon appear in his momentous paper, it is worth briefly exploring one other concept that arose from this work, which would not have such direct implications, but would totally transform the nature of physics – a remarkable feat, given Faraday's lack of formal education.

In trying to explain how the magnet could influence the wire remotely to induce a current, Faraday came up with the idea of lines of force – the constituents of what is known as a field. If you've ever played with a bar magnet beneath a sheet of paper with iron filings on it, you might have seen the way that the metal fragments are assembled by the magnetism into a series of curved lines stretching from pole to pole. Faraday imagined that when an electromagnet starts up, these lines move out into position, like an umbrella unfurling. As they did, it was premised, they would cut across the wire in which the current was being induced. And, Faraday suggested, it was the cutting of the lines of force – whether by switching on and off an electromagnet or moving a permanent magnet – that induced the electrical current.

The lines of force were developed theoretically into the concept of a field – a phenomenon that had values at every point in space, values that could change with time. This 'field' concept is at the heart of most modern physics.

With these ideas whirling in his mind, Faraday pulled together his thoughts on that Thursday in November, not in his workplace but at the country's senior scientific organisation, the Royal Society.

Experimental Researches: the 1831 day

Faraday's paper, 'Experimental Researches in Electricity', was read at the Royal Society on 24 November 1831 and published in the *Philosophical Transactions* the next year. In it, Faraday set out the nature of electrical induction, the generation of electricity from magnetism, covering what he described in typical Victorian fashion as 'a new Electrical Condition of Matter' and 'Arago's Magnetic Phenomena'.

This wasn't the first time that induction – the ability to produce an electrical current in one conductor that is near to but not in direct contact with another conductor carrying electricity – was mentioned. But what Faraday did was to open up a new field that had only been skirted around until this point, making possible the development of a whole new application of electrical generation and electric motors. As he put it, 'the hope of obtaining electricity from ordinary magnetism [has] stimulated me at various times to investigate experimentally the inductive effect of electric currents.'

Faraday systematically took his audience through a series of experiments, detailing precisely how he went about them. For example: 'About twenty-six feet of copper wire one twentieth of an inch in diameter were wound round a cylinder of wood as a helix, the different spires of which were prevented from touching by a thin interposed twine. This helix was covered with calico, and then a second wire applied in the same manner. In this way twelve helices were superposed, each containing an average length of wire of twenty-seven feet, and all in the same direction. The first, third, fifth, seventh, ninth, and eleventh of these helices were connected at their extremities end to end, so as to form one helix; the others were connected in a similar manner; and thus two principal helices were produced, closely interposed, having the same direction, not touching anywhere, and each containing one hundred and fifty-five feet in length of wire.'

One helix was connected to a galvanometer – an instrument for measuring the presence of an electrical current – the other to a battery. In this case, with no movement, as Faraday put it, 'not the slightest sensible deflection of the galvanometer needle could be observed'. But he persevered and noticed a small effect in the galvanometer circuit when the current was switched on and off and a much larger one when a zigzag of wire with a current running through was moved towards and away from a second such circuit for the galvanometer. When movement stopped, so did the effect.

It's telling of the uncertainty of the consistent nature of electricity at the time that Faraday noted that he could find 'no evidence by the tongue,* by spark, or by heating fine wire or charcoal' – or for that matter chemical effects produced by the induced current. He suggested that 'this deficiency of effect' is not because the induced current can't pass through fluids – which would imply some different form of electricity from static electricity – but probably because of its brief duration and feebleness.

Similarly, he tried out a similar experiment using 'normal electricity'. Again, this was static electricity, using a Leiden jar, a means of storing an electrical charge – a clumsy version of what's now called

---

* This is not an obscure instrument, but literally what it says. The human tongue is quite sensitive to an electrical current.

a capacitor. Rather than produce a steady flow of current like a battery, this would have produced a very rapid one-off surge, which, as Faraday noted, made it near-impossible to separate the two effects where a flow of electricity started and finished. Again, there was yet to be certainty that such normal electricity and what he described as 'voltaic electricity' – the electricity from a battery – were the same thing, but operating in a different way.

A painstaking exploration

Faraday then moved on to look at the 'evolution of electricity from magnetism' – a natural follow-on from the first series of experiments given what was already known about electromagnets. Rather than have a wire acting on another wire, he wrapped the wires around an iron ring, turning it into an electromagnet which again produced brief induced currents when switched on and when switched off. He then produced similar effects with 'ordinary magnets' – simple bar magnets.

Reading the paper now emphasises Faraday's painstaking doggedness. Experiment after experiment – over a hundred in total – was undertaken with various configurations and materials. He next took a dead-end route to describe something he refers to as the 'electro-tonic' state, a theory that the matter in a wire subject to induction is put into a peculiar state – though in a footnote, he noted that 'later investigations of the laws governing these phenomena induce [pun intended?] me to think that the latter can be fully explained without admitting the electro-tonic state'. He later dropped the concept, realising that what he had felt was a separate state was only a reflection of the way that the magnetic lines of force operated.

Finally, Faraday used the concept of induction to explain what was happening in Arago's 'magnetic phenomena' mentioned above, where a copper disc dragged a magnet along with it despite copper not being a magnetic material. Faraday realised that the relative motion of the magnet and the disc could induce a current with the copper, which then had an electromagnetic effect. Again, the paper takes us into the detail, for example:

> The galvanometer was roughly made, yet sufficiently delicate in its indications. The wire was of copper covered with silk, and made sixteen or eighteen convolutions. Two sewing-needles were magnetized and passed through a stem of dried grass parallel to each other, but in opposite directions, and about half an inch apart; this system was suspended by a fibre of unspun silk, so that the lower needle should be between the convolutions of the multiplier, and the upper above them.

Faraday's paper gave an insight into his working methods and ability to take on a challenge through meticulous detailed experimentation. But it also marked the beginning of a new world – one where electricity would be transformed from an entertaining parlour trick to the power source of everyday life. When Faraday read this paper, he would have done so by gas light. Within a few decades, electricity, generated by the induction effect, would be taking over the world.

By the next year, Faraday and others were producing crude electrical generators, the same year that a practical electric motor was first demonstrated. An electric train would be demonstrated as early as 1837, though this was battery powered. It would take a few more decades, to the 1870s, for electric trains to become a commercial proposition. Electric arc lighting, powered by generators, would also be introduced in the 1870s, soon followed by incandescent bulbs.

Faraday, the person

During Faraday's lifetime, the role of scientist moved from the domain of the gifted amateur to a professional career. The very word 'scientist' was not coined until three years after the reading of his paper. At the time he would have been known as a natural philosopher, a term that was in part superseded because the 'real' philosophers felt that people like Faraday were not worthy of the title.

Faraday had no university education, and, in contrast with a modern physicist whose work is likely to be dominated by maths, he hardly ever used more than arithmetic. Although Newton had been mathematically oriented, this is not entirely surprising, as Newton thought of himself as a mathematician. Other physicists of Faraday's time may have had a stronger academic background, but many of them had limited mathematical experience. When James Clerk Maxwell (see Day 4) published his purely mathematical work on electromagnetism in the 1860s, many, including leading lights of the day such as William Thomson (Lord Kelvin) admitted that they struggled with it.

Faraday held strongly to his religious beliefs, which included a literal interpretation of the Bible. Although a practised public speaker, he generally avoided socialising (though he enjoyed music and the theatre), preferring the company of his family. He apparently had some enthusiasm for the velocipede, an early form of bicycle. He was regarded as a gentle and kind man, though his successor at the Royal Institution, John Tyndall, pointed out that it was important not to limit him to a caricature. 'Underneath his sweetness and gentleness,' Tyndall wrote, 'was the heat of a volcano. He was a man of excitable and fiery nature; but through high self-discipline he had converted the fire into a central glow and motive power of life, instead of permitting it to waste itself in useless passion.'

Life changers

Generators

The immediate development from the 1831 paper were devices for creating current electricity. Initially these were dynamos where a coil of wire was rotated in a magnetic field, inducing a current. Later, alternators became more common, producing the alternating current that is the main way that grid electricity has been used since the early 20th century. Typically, these will involve a moving magnet with a static coil, but still rely on Faraday's discoveries.

Transformers

One of the reasons that alternating current (AC) proved popular is that transformers, changing the voltage up or down, are far easier to produce with this type of current than with direct current, which constantly flows in the same direction. Because the AC current is always changing, a coil carrying such a current will continuously induce a current in another coil – by varying the number of windings in the two coils, different voltages can be produced, all dependent on Faraday's discovery.

Wireless charging

Increasingly phones, electric toothbrushes and other battery devices are charged without wires being plugged into them by being placed on a wireless charger. Such chargers induce a current in a coil in the device to be charged, again based on Faraday's discovery.

Monday, 18 February 1850

Rudolf Clausius – Publication of 'On the Moving Force of Heat'

He may not be as familiar as Newton, but German physicist Rudolf Clausius was, nevertheless, a major player in the development of the science of thermodynamics. On this day, his paper that would establish the second law of thermodynamics was read to the Berlin Academy. This law is fundamental to understanding the flow of heat and the working of engines dependent on heat – the second law is even considered to be the aspect of nature that drives our idea of the progress of time. Clausius is a scientist whose name deserves to be better known for a principle that can be seen in action across the world.

The year 1850

Familiar names were formalised this year: American Express was founded, while both Los Angeles and San Francisco were incorporated as cities, shortly before California was admitted as a US state. The poet William Wordsworth died, and Robert Louis Stevenson was born, while US vice president Millard Fillmore became the thirteenth president on the death of Zachary Taylor. Australia got its first university, the University of Sydney.

Clausius in a nutshell

Physicist and mathematician

Legacy: thermodynamics and entropy

Born 2 January 1822 in Köslin, Prussia (now Koszalin, Poland)

Educated: Universities of Berlin and Halle

Professor of physics, Berlin, 1850–55

Professor of physics, ETH (Eidgenössische Technische Hochschule – Federal Institute of Technology), Zürich, 1855–67

Married Adelheid Rimpam, 1859

Professor of physics, Würzburg, 1867–69

Professor of physics, Bonn, 1869–88

Organised ambulance corps in Franco-Prussian War – wounded in battle, 1870

Married Sophie Sack, 1886

Died 24 August 1888 in Bonn, Prussia, aged 66

Rudolf Clausius

The mystery of heat

Like many physicists before the 20th century, Rudolf Clausius did not concentrate on a single topic. His early work was on the colour of the sky. Unfortunately, the approach he took, assuming that the blue of the daytime sky and the redness in the vicinity of the setting Sun was due to reflection and refraction was wrong.

The correct explanation would not be made until 1899, when John Strutt (Lord Rayleigh) showed that light was being scattered by atmospheric molecules, with blue light more likely to be diverted off course by interaction with gas molecules in the atmosphere. Blue light was therefore spreading across the sky, while the light towards the red end of the spectrum remains near the Sun – especially at sunrise and sunset, when the light has more air to get through due to passing into the atmosphere at a shallow angle.

By a coincidence, the next field that Clausius moved on to was also one where a misunderstanding of what was happening had resulted in an incorrect model – though in this case, the invalid science had nonetheless allowed useful deductions to be made. In the 18th century it had become widely accepted that heat was an invisible, intangible fluid that flowed from hot objects to cold ones. This fluid, given the name 'caloric' by French chemist Antoine Lavoisier, was thought to be conserved – not created nor destroyed, but flowing from body to body when they were in contact.

The French engineer Sadi Carnot effectively started the science of thermodynamics – the study of the flow of heat – which was essential to understand the efficient workings of the increasingly important steam engine. Carnot wrote a book *Réflexions sur la Puissance Motrice du Feu* (Reflections on the Motive Power of Fire) in 1824, shortly after Clausius was born, which explained the working of steam engines as being due to the transfer of caloric from a hot body to a colder one.

Sadly, Carnot died at the young age of 36 in 1832 and his work was not widely read, but his ideas were spread by an 1834 paper by another French engineer, Émile Clapeyron. By this time, the caloric theory that Carnot had based his thinking on was falling into disrepute.

The first attack on caloric came significantly before Carnot's work in 1798 when Count Rumford made an experimental test of its nature. Rumford was a colourful figure – an American-born Englishman who was knighted in the UK and made a Count of the Holy Roman Empire in recognition of his later work in Bavaria. It was while there that his observations of cannon construction led him to question the existence of caloric.

The barrels of cannons were produced by boring out a solid metal cylinder. As anyone who has felt a drill bit after using it knows, the process of boring a hole generates a considerable amount of heat as a result of friction between the drill and the material. Rumford made use of a particularly blunt boring bit, drilling into a cannon blank that was immersed in water, measuring the increase in the temperature of the water, which could be brought to boiling point by the intense friction.

Bearing in mind that caloric was supposed to be contained in an object and was theoretically conserved, the result of constantly boring the cannon should have been to drain it of caloric. But Rumford found that the supposed caloric appeared to be inexhaustible – heat was generated as long as he was able to go on boring. Rumford deduced that heat was in some way connected to movement, initiated by the friction. His work was extended by a different kind of physicist – the Manchester-based brewer, James Joule.

Joule, who was interested in the potential of improving the technology in the family brewery by moving from steam power to Faraday's electrical motors, measured both the heat generated by electricity and that produced by mechanical work, quantifying the relationship using, among other devices, an experiment that linked a falling weight to a rotating paddle in a container of water, measuring the increase in temperature produced.

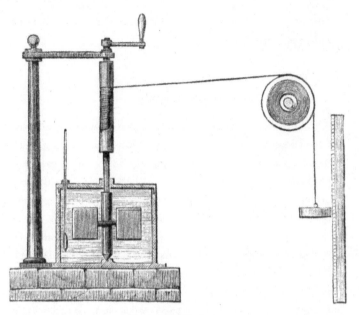

Joule's paddle device.

By the late 1840s, it was becoming clear that heat was a form of energy – and that it was energy, rather than the non-existent caloric, that was conserved.

Ditching the caloric: the 1850 day

Although caloric was on its last legs thanks to the work of Rumford, Joule and others, it was the publication of Clausius' 1850 paper 'On the Moving Force of Heat' that finished it off entirely. Because caloric was considered a substance in its own right, Carnot and other supporters of caloric theory assumed that the heat in a substance reflected the nature of the substance itself. Clausius dismissed this, making it clear that the maximum amount of work that could be produced from heat was purely dependent on the absolute temperatures of the heat reservoirs involved. The type of material used had no impact.

Heat reservoirs

If we think of a simple heat engine, such as a steam engine, it's easy to think of it using the energy released by burning the fuel to boil the water, which enables the steam to push a piston and power the engine. However, this picture misses out on an essential part of any heat engine – a colder part. Heat engines work by heat moving from a warmer part to a colder part, the latter often referred to as a 'cold sink', doing work in the process.

In a steam engine, for example, the piston is forced in one direction by expanding steam, but then has to return, which it does as a result of cooling. Steam engines achieve this either by venting steam into the atmosphere or using a condenser, which can be as simple as a jacket of cold water. A large-scale example of a condenser being used to provide a cold sink is in the cooling towers used at power stations to cool water that has been used in steam turbines.

Clausius showed, based on Carnot's work but without the requirement for caloric, that it was the difference in absolute temperatures (temperature above the absolute minimum temperature of −273.15°C [−459.67°F]) between the hot and cold heat reservoirs that was the only determinant of the maximum possible efficiency of a heat engine.

This wasn't the only blow that Clausius landed on caloric theory in his 1850 paper. The other requirement of that theory was that the heat in a system is conserved. Caloric could not be created or destroyed, it just flowed from place to place. As Rumford and Joule had shown, this wasn't the case, and Clausius instead developed the first law of thermodynamics in the form that when work is done by heat, the latter is converted to the former. It is not heat, but energy that is conserved. It can be translated from one type to another, as demonstrated by Joule, but the total remains the same. (By 1905, when the paper we meet on Day 6 was written, it would become clear that even this is an oversimplification. It is not energy that is conserved but mass–energy. Just as heat and work were shown to be interconvertible, so would matter and energy be.)

It's important to realise that at the time of writing this paper, Clausius was not a well-established physics professor with an international following. He did not receive his doctorate until the summer of 1848, which was for his incorrect explanation of why the sky was blue and the sunset red. Clausius went on to produce his theory of heat paper while still at the University of Halle shortly after his doctorate. It was as a result of the paper that he won his first major academic post as professor of physics at the Royal Artillery and Engineering School in Berlin, with teaching rights at the University of Berlin.

Second law

The second law of thermodynamics, which was first clearly stated in Clausius' paper on moving heat, is one of those apparently simple things that in reality does much more than might be expected. Heat became a speciality for Clausius, and, as the name suggests, the second law is about the movement (dynamics) of heat. The way that Clausius phrased the second law, translated into English, was 'Heat can never pass from a colder to a warmer body without some other connected change occurring at the same time'.

Since Clausius' time, two other laws have been added. The zeroth law (so-called because it is technically more fundamental than the first law) dates to the 1930s, and effectively fills a potential loophole by saying that if two systems are both in thermal equilibrium with a third system (where there is no net heat flow between them), then they are in thermal equilibrium with each other. And the third law, from the early 20th century, deals with a situation that is unlikely in the natural world, stating effectively that it is impossible to reduce the temperature of a body to absolute zero in a finite number of steps.

At first sight, the second law feels trivial. Surely it is obvious that heat goes from warmer to colder bodies, rather than the other way round? But in physics, the apparently obvious, common-sense view is not always correct and always needs proving. One obvious problem is presented by the existence of the refrigerator. This takes heat out of its cool interior and pumps it into the warmer room around it – exactly the opposite of the prediction made in the law.

However, the second law only applies if there is no energy going into the system. Refrigerators don't spontaneously move heat from a cold place to a warm place – it takes energy to make it happen. Once we can put energy into a system, it's perfectly possible to run such a heat exchanger. Another circumstance in which an external source of energy results in a reversal of the second law (in the form of entropy, as described below) is when the external source of the Sun pumps energy into the Earth.

Pump up the entropy

The initial statement of the second law was all about the movement of heat, but Clausius was the first to realise that something more was involved, a something that he would name 'entropy' (*die Entropie* in German). His intention in coming up with the name was to draw a parallel with the word 'energy' (*die Energie*). Just as Clausius understood energy to represent the work content of something, so entropy represented what he would refer to as 'transformation content'.

What Clausius was reflecting was a concept that went back to Sadi Carnot, that some heat would always be lost to the environment without producing useful output when heat was converted into work – there could not be a 100 per cent efficient machine. The distinction Clausius had in mind between work content and transformation content was that some percentage of a quantity of heat would produce work, but some would be used up in the transformation process.

Through the 1850s Clausius would refine this concept until he reached the first entropy-based version of the second law in 1862, where he stated that the sum of the transformations (change in entropy) in a system could only be positive, or at a minimum, zero. To put it another way, the entropy in a closed system is expected to stay the same or to rise.

In a sense, Clausius' introduction of the second law was ahead of its time. He realised that heat was related to the kinetic energy of movement in the component parts of a body, but would not take it quite as far as some of his younger contemporaries. It was only in the 1870s that the statistical mechanics view, pioneered by Ludwig Boltzmann and James Clerk Maxwell, made it clear just what entropy really was. This approach saw the heat in a system as the sum of the energies of the particles in a substance. So, for example, in a box of gas, the heat was a result of the way that the gas molecules moved around the box at high speed – they had the kinetic energy of motion which defined the heat present.

Seen through this clearer perspective, entropy was transformed into a measure of the number of ways the components of a system could be organised. The more ways that were possible, the higher the entropy. How this leads to the second law could be seen from a simple case of having two boxes of gas – one hot, one cold – with a partition dividing them.

In this setup there is relative order. Of course, there are many different ways that the individual gas molecules could be positioned in each box, but we know that, say, the faster moving molecules are all in the box on the left and the slower moving molecules in the box on the right. However, if we open the partition so the gas molecules can freely move between the two boxes, over time we would expect to find a mix of hot and cold molecules in each box. There are many more ways of organising the mix of molecules than when we have hot molecules in one place and cold in the other. So, the entropy has been increased.

In the old thermodynamic terms, heat has moved from the hotter box to the colder box. But for this process to run in reverse, we need the unlikely outcome that the hot molecules would, of their own accord, head in one direction and the cold molecules in the other, so the heat would end up moving from a cooler box to a warmer box. This is a very unlikely outcome to happen to any great degree.

Note, though, something with which Clausius would have been uncomfortable. In this new formulation, the second law has become statistical, not absolute. It's a good bet, not a certainty. Although in any sensible circumstance we would imagine that entropy will stay the same or increase, it is possible for entropy to spontaneously reduce. This feels unnatural. Bearing in mind entropy is a measure of the disorder in the system, it's like expecting a broken egg to unbreak of its own accord. However, given enough time, statistically we would expect it occasionally to be the case that entropy would spontaneously reduce. However, there are so many molecules in even a small box of gas that it might require billions of years for anything more than a brief, tiny reduction to occur.

Maxwell's demon

The statistical nature of the second law of thermodynamics was illustrated neatly by Scottish physicist James Clerk Maxwell, using an imaginary being that became known as Maxwell's demon.

As above, we have an experiment with two boxes of gas, but start with them at the same temperature. In practice, not all molecules will be travelling at the same speed. Some will be quicker and some slower: when we say they are at the same temperature, we mean the average velocity of molecules in each box is the same. Maxwell imagined placing a tiny being in charge of a door between the two boxes. If a fast molecule is heading from left to right, the demon opens the door and lets it through. The same goes for a slow molecule heading from right to left. But other molecules are not allowed through.

As a result, heat will move from cold to hot. The hotter side will get hotter and the cold side colder. But no work is being done on the system (the door is allowed to be frictionless). This appears to run counter to the second law. In reality, it's more of an illustration of the statistical nature of the law. Although various attempts have been made over the years to show that the demon couldn't do its job, the concept has never been entirely disproved. The demon remains an entertaining sideshow of the second law.

Clausius, the person

Brought up in a large family, Clausius was educated at his father's school (the elder Clausius was a school principal and a church minister) before moving on to the Stettin Gymnasium (high school). On going to the University of Berlin, his first interest was history, but he became increasingly interested in mathematics and physics, graduated and went on to undertake further academic positions in physics.

After the early incorrect work on light, his 1850 paper was his first major publication, yet remains his most famous work. He would continue to work on heat until the mid-1870s, when he switched his focus to electromagnetism. His work would be interrupted by the Franco-Prussian war between 1870 and 1871. Despite already being nearly 50, Clausius led an ambulance corps of students from Bonn University and was injured in the field, receiving the Iron Cross.

Clausius was married twice. His first wife, Adelheid, died in 1875, giving birth to the sixth of Clausius' seven children (though only four would survive into adulthood). He remarried in 1886 to Sophie. Clausius was a determined worker, still said to have been continuing his academic work on his deathbed.

The patriotism that led Clausius to support the war also seems to have blinded him sometimes to the advantages of international cooperation in science. He resented the suggestion that Joule had made discoveries ahead of Clausius' German compatriot Julius von Mayer, and he challenged the originality of the work on heat of the Scottish physicist James Clerk Maxwell, even though Maxwell had been scrupulous about acknowledging the aspects of his work that were built on papers Clausius had written. Nonetheless, Clausius was comfortable with accepting a Fellowship of the Royal Society of London in 1868.

Life changers

Internal combustion engines

Although by the mid-21st century the internal combustion engine is likely to have become pretty much extinct, for over 100 years it was an essential driver of the development of technological civilisation. Developed into a working device in the 1870s, the internal combustion engine, powered by petrol or diesel, might have seemed very different from steam engines, but both were heat engines, making use of the physics that Clausius pioneered to transform the generation of heat into motive power.

Power stations

The English language is particularly weak on energy and power. We speak of a power station generating energy, but in reality, a power station is a mechanism for transforming energy from one form to another, obeying the first and second laws of thermodynamics. Apart from nuclear, pretty well all the energy for a power station originates from the Sun as light, which will in one or more stages be converted into electricity. (Even power stations that burn fossil fuels are using light energy that has been stored as chemical energy in plants, which is released in combustion to produce heat which drives a generator to produce electricity.)

Heating systems

Heating is such a basic need – at its most basic provided by lighting a fire – that it can be hard to remember that it is also a thermodynamic process. The first law is involved, in that heating systems often involve the transfer of energy from one form to another, while the second law ensures that by making a radiator or similar device warmer than the surrounding air, heat will move from the radiator out into the room.

Fridges and air con

Refrigerators and air conditioning units provide classic demonstrations of the second law of thermodynamics in action. They are effectively heat pumps, transferring heat from one place (the inside of the fridge or the room) to another. Air conditioners typically move the heat outside the building, while a fridge has a radiator on the back that sends heat into the room. This is only possible because of the energy put into the device, usually from electricity.

Monday, 11 March 1861

*James Clerk Maxwell – Publication of
'On Physical Lines of Force'*

Scottish scientist James Clerk Maxwell had wide-ranging interests, from colour vision to the kinetic theory of gases – but his biggest contribution to the modern world was his bringing together electricity and magnetism in a series of equations that would drive the future of electromagnetic applications and predict the existence of radio waves. He revealed his advanced ideas on electromagnetism to the world on this day – ideas that now are central to the understanding of electromagnetism. It's not for nothing that Richard Feynman said: 'From a long view of the history of mankind – seen from, say, ten thousand years from now – there can be little doubt that the most significant event of the nineteenth century will be judged as Maxwell's discovery of the laws of electrodynamics.' A kind, gentle individual with a distinctive sense of humour, Maxwell is the greatest physicist most people have never heard of.

The year 1861

Soon after Kansas became the 34th state, the American Civil War broke out in 1861, following the election of Abraham Lincoln as sixteenth president of the United States of America. The Kingdom of Italy was declared prior to completion of reunification in 1871. The world's first full iron-hulled battleship, the HMS *Warrior* was commissioned in the UK. The University of Washington was founded. The controversial educator Rudolf Steiner, military men Edmund Allenby and Maximilian von Spee, first prime minister of Iceland Hannes Hafstein and pioneering filmmaker Georges Méliès were born, while King Frederick William IV of Prussia, poet Elizabeth Barrett Browning, gunsmith Eliphalet Remington and Queen Victoria's husband Prince Albert died.

Maxwell in a nutshell

Physicist

Legacy: colour theory, kinetic theory of gasses and electromagnetism

Born 13 June 1831 in Edinburgh, Scotland

Educated: Universities of Edinburgh and Cambridge

Fellow of Trinity College, Cambridge, 1855

Inherited the Glenlair Estate in Dumfries and Galloway, 1856

James Clerk Maxwell

Professor of natural philosophy, Marischal College, Aberdeen, 1856–60

Married Katherine Dewar, 1858

Professor of natural philosophy, King's College, London, 1860–65

Cavendish Professor of experimental physics, University of Cambridge, 1871–79

Died 5 November 1879 in Cambridge, England, aged 48

The power of analogy

We have already seen the birth of Faraday's descriptive theory of electromagnetism on Day 2, but Maxwell would uncover far more of this phenomenon. Even so, Faraday's concept of fields is absolutely central to modern physics and, crucially, it would give Maxwell the inspiration to come up with his 1861 paper – those lines of force even appear in the title. This paper would not only provide the basis of modern electromagnetic devices; it would enable Maxwell to understand what light was and would make possible the development of radio, radar, microwave ovens and other technology based on electromagnetic waves.

James Clerk Maxwell might now seem something of a prodigy. He became a professor of natural philosophy (the most common term for what we would now call science) at the age of 25, when a modern physicist would probably have only just completed her doctorate. Yet at the time, such rapid promotion was not unheard of. Two men who remained close friends of Maxwell's throughout his adult life, William Thomson and Peter Tait, had become professors younger – Thomson became Professor of Natural Philosophy at Glasgow University when he was only 22, while Tait became Professor of Mathematics at Queen's College, Belfast when he was 23.

Unlike Faraday, Maxwell had a privileged upbringing: brought up on his family's country estate, he attended the best universities. When he inherited the estate, he could have dedicated his time to running it. But his fascination with how things worked would not let him go, and he proved to have an instinctive grasp for the relationship between mathematics and the real world.

Maxwell had a number of interests he returned to throughout his working life. At the time, the most widely recognised of these was his work on statistical mechanics, showing how the combined action of many molecules could be used to predict the behaviour of gases, for example, and contributing to the understanding of the second law of thermodynamics we discovered in the previous chapter. But from today's perspective, without doubt, it was his work on electromagnetism that made Maxwell such an outstanding physicist.

In his first paper on the topic, written at the end of 1855 and published in 1856, Maxwell made it clear where his concepts came from – it was entitled 'On Faraday's Lines of Force'. In it, Maxwell says, 'In order to obtain physical ideas without adopting a physical theory we must make ourselves familiar with the existence of physical analogies.' What he meant by this was that because there often seemed to be similarities between physical laws, if there was something you did not understand, you might be able to at least partially explain it based on what was already known.

Building a model

Maxwell's 'physical analogies' were a part-way step to what is described as modelling. Modelling in the scientific sense is constructing what is usually a simplified version of reality. Initially such models were often actual physical objects. Maxwell himself, for example, had built a model to help understand the interactions of the components of the rings of Saturn. However, Maxwell's analogies were theoretical descriptions of other, known systems such as fluid flows or mechanical structures, which were thought to behave in a similar way to the field being studied.

Maxwell's biggest breakthrough was in realising that those mathematical models did not have to be based on any known physical situation. Modern physicists, since Maxwell, try to get a better understanding of the world around them by building systems of mathematics that produce numbers that correspond to what happens in the natural world.

In the 1856 paper, Maxwell used the analogy of electricity as being like a fluid flowing through a porous substance, while magnetism was like vortices, swirls, that built up within the fluid. The fluid flows corresponded to Faraday's lines of force. (The terminology we have inherited from the period still tends to refer to electricity as if it were a fluid flow – words like electrical 'current', for example, while the vacuum tubes used in early electronics were called valves in the UK.)

This first model gave some indication of success – it matched several of the behaviours of electricity and magnetism. But Maxwell was clear that it was not intended to be the electromagnetic equivalent of caloric. There was no suggestion of a real electrical fluid existing. He commented, 'I do not think that [the fluid analogy] contains even the shadow of a true physical theory; in fact, its chief merit as a temporary instrument of research is that it does not, even in appearance, account for anything.' Having got his model partly working, Maxwell put the problem aside for some time, but he would come back to it in 1861 with a much more powerful analogy.

Maxwell's marvellous mechanical models: the 1861 day

In moving from 'On Faraday's Lines of Force' to the publication of his 1861 paper 'On Physical Lines of Force' (titles which are sufficiently similar that the two are often confused in referencing them), Maxwell moved to a more solidly mechanical model of electromagnetism. The fluid model was limited because it only worked for fields that did not move, which was extremely limiting when considering practical applications of electricity such as generators and motors, where it was the movement of or through fields that made things happen.

Once again, this was a scientific model, based on the analogous behaviour of mechanical objects. After some initial work on a model using spheres that expanded as they spun around, Maxwell settled on an elegant model which involved a series of a rotating hexagonal cells, each of which was surrounded by large numbers of small objects like the ball bearings that support a rotating joint. The cells he referred to as vortices and the ball bearings as idle wheels.

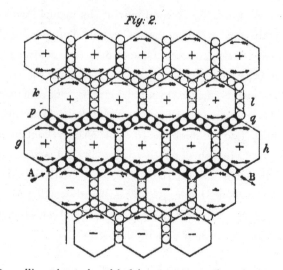

Fig: 2.

Maxwell's mechanical model of electromagnetism from the 1861 paper.

When an electric current was applied, the ball bearings started to flow through the system representing that current, which caused the hexagonal cells to rotate – that rotation represented the magnetic field produced by the flow of electricity. This model added the important induction mechanism to the analogy, as the reaction between the ball bearings and the rotating hexagonal cells would produce a temporary flow in a second layer of ball bearings when the current was switched on or off.

Although this was still an analogy, Maxwell believed that he was now dealing with something closer to reality. At the time it was widely believed that all space was filled with a material called the luminiferous aether. The archaic spelling is now generally rendered 'ether', which is how we will refer to it from now, but to be clear, this is not related to the organic compound ether used as an early anaesthetic. (This liquid was named after the conceptual ether because of its volatility.)

The existence of the ether was proposed because light was known to be a wave, and all the other waves we know of are disruptions that pass through a medium. But light traverses the vacuum of space, which suggested that there was something else, something invisible, out there in which light could wave. Maxwell suspected that magnetic effects included vortices (hence his term) in the ether, so that he believed this model to be significantly closer to reality than his fluid model.

He was less certain about the electrical component of his model. He commented: 'The conception of a particle having its motion connected with that of a vortex by perfect rolling contact may appear somewhat awkward. I do not bring it forward as a mode of connexion existing in nature ...' Nonetheless, the model worked well, so his suspicion was that there was something in nature that was being represented by the ball bearings.

Electromagnetism and light

Two key aspects of Maxwell's work on electromagnetism came soon after the publication of his paper (or, strictly, his three papers, as the work appeared in three parts). To deal with an aspect of the behaviour of electromagnetism his model did not yet cope with, Maxwell tried making the hexagonal cells have some elasticity, meaning the cells could twist and contract, which enabled him to add in the effect of the electrostatic attraction that occurs between two opposite electrical charges.

Although it wasn't Maxwell's intention, adding in this extra feature had a mind-bending implication. If a material is elastic, it is possible to send a wave through it. You can't have a wave travelling through something that is totally rigid, as the whole nature of a wave is to have an oscillating movement passing through the medium. In effect, a twitch in one level of ball bearings in his updated model would twist the adjacent cells, which would twitch the next layer of ball bearings and so on. Thinking of what the model was representing, a changing electrical field would produce a changing magnetic field, which would produce a changing electrical field and so on.

No more ether

Maxwell decided his electromagnetic wave would need nothing more than the ether to progress through empty space. In reality, though, he didn't go far enough. Maxwell's hero Michael Faraday had suggested back in 1846, when Maxwell was just fifteen years old, that a wave travelling through fields would not need the presence of the ether. Faraday had said:

> The views which I am so bold as to put forth consider, therefore, radiation as a high species of vibration in the lines of force which are known to connect particles, and also masses of matter, together. It endeavours to dismiss the ether, but not the vibrations.

Faraday would be proved entirely correct in this regard. A series of experiments carried out by American physicists Albert Michelson and Edward Morley from 1887 (see more on this on Day 6) could not detect any effect of the Earth moving through the ether, which would be expected if it existed. At the start of the 20th century, Albert Einstein went on to show that the ether could not be sensibly supported as a concept.

There was already a wave that was known of that could fit the bill of Maxwell's theoretical electromagnetic wave capable of travelling through the vacuum of space: light. And unlike Faraday, in his speculation on such a wave, Maxwell had the additional weapon of mathematics in his armoury. His model enabled him to calculate the speed such a wave would have to move at in order to exist. Maxwell calculated that the wave would travel at 193,088 miles per second (310,745 kilometres per second). He suspected this was around the right speed for light, but he had a problem no modern physicist would face. He was on his summer break at his Scottish home, far from the library and his journals at the university in London. He had no way to check the best modern measurements of light speed.

It was only when Maxwell returned to London months later that he was able to access the figures and discovered that his estimate for the speed of his electromagnetic waves was within 1.5 per cent of the speed of light as then measured. He would extend the 1861 paper the next year to include the mechanism behind this, known as displacement current, and the implication of electromagnetic waves.

The second development, which Maxwell would complete in 1864, was to move away from his mechanical model to an entirely mathematical one. This was a new way of thinking: many of the great physicists of the day struggled with Maxwell's purely mathematical view, where there was no real-world analogy, just a representation of what was happening in the form of mathematical equations. Maxwell likened it to having a set of church bells being operated unseen from below. All the ringers saw was a set of ropes. Without having any idea of what was happening above the ceiling of the ringing chamber, it would be possible to describe the movement of the ropes using mathematical formulae. It did the job without ever having any model of the behaviour of the bells.

Maxwell produced a series of twenty equations, between them mathematically summarising the behaviour of electricity and magnetism. The central twelve of these would be combined and simplified by English electrical engineer Oliver Heaviside in 1884 as four powerful, simple-looking equations, usually described simply as Maxwell's equations.

Maxwell, the person

Maxwell could so easily have been a dilettante amateur scientist, dabbling without ever achieving anything significant. He developed an interest in science early in life, inspired in part by the wild landscape around his home. As a child he would often ask, 'Show me how it doos', and 'What's the go o' that?' He built a home laboratory, always experimenting with the materials he had to hand. But rather than fall back on his laird of the manor position when he inherited the estate, he pushed forward with an academic career that would last the rest of his short life.

Although Maxwell had a privileged upbringing, his parents allowed him to mix with the local farm children, and this seems to have inspired a lifelong enthusiasm for the education of those from humbler backgrounds: at each of his university positions he was involved in the educational programmes of working men's institutes. This also reflected his strong Christian ethos.

Maxwell was both mathematically visionary and able to make unexpected leaps in his scientific work – and there seems to have been very little division between his work and home life. He wrote thousands of letters to scientific friends that would veer suddenly between social remarks and the exploration of the latest physics. Until his last post, when he set up the world-leading Cavendish Laboratory in Cambridge, he had very limited experimental facilities in the universities where he worked, and so did a considerable amount of experimental work in his homes, aided by his wife, Katherine.

Though work was so central to Maxwell's life, it would be a poor picture of the man if that was all that was considered. He was initially socially awkward, but developed strong friendships, bolstered by an ever-present sense of humour. His letters are littered with jokes, and they would even creep into his serious business. So, for example, when listing the facilities that he hoped to provide in the Cavendish Laboratory in a letter to his friend William Thomson (later Lord Kelvin), Maxwell specified 'A gas engine (if we can get it) to drive apparatus, if not, the University [boat] crew in good training in four relays of two, or two of four according to the nature of the expt.'

Maxwell was also a lifelong poet, whether writing poems inspired by the tedium of working through problems when a student, his feelings for Katherine, or the scientific developments of the day. Maxwell was no stuffy, two-dimensional Victorian, but a rounded individual.

Life changers

Electromagnetic devices

While Faraday and his contemporaries made basic electromagnetic devices such as motors and generators possible, it was only with Maxwell's theoretical underpinnings for electromagnetism that it would be possible to develop the whole range of electrical and electronic devices our modern society depends on.

Radio/microwaves/TV/X-rays

Radio was the first part of the electromagnetic spectrum to be demonstrated to work in the same way that Maxwell predicted, by the German scientist Heinrich Hertz, relatively soon after Maxwell's death. The understanding of electromagnetic waves would go on to be expanded to take in a much wider use of other parts of the electromagnetic spectrum, such as the high-energy X-rays, or the radiation at the high-frequency end of the radio spectrum that would become known as microwaves.

Mobile phones

Arguably, the mobile phone shows the greatest benefit from Maxwell's legacy, combining as it does the vast range of electronic components with the use of electromagnetic waves in the cell phone's radio system.

Einstein's inspiration

Although not a practical use, Maxwell's work was also a life changer in acting as inspiration for Albert Einstein. It was Maxwell's discovery of the fixed speed of light that was the key to Einstein developing the special theory of relativity. Einstein had a picture of Maxwell on his study wall, and described Maxwell's breakthrough of moving to a mathematical description of fields as 'the most profound and the most fruitful [change] that physics has experienced since the time of Newton'.

Monday, 26 December 1898

Marie Curie – Publication of 'On a New, Strongly Radio-active Substance'

All the more remarkable for a woman at a time when the concept of sexual equality hardly existed, Marie Curie was the first person to win two Nobel Prizes. Her work, first with husband Pierre and then alone after his death, would take radioactivity from an obscure curiosity to something that was understood as both useful and dangerous – eventually her exposure to radioactive materials and X-rays would end her life. Curie's discovery of the radioactive element polonium was significant, but her breakthrough paper in 1898 told of the far more significant discovery of radium. Although Curie did not discover X-rays, she helped bring them into widespread medical use, transforming their application, and she introduced radiotherapy to the medical world.

The year 1898

In 1898, a number of constituent areas were consolidated to form the modern city of New York. In England, the first person was killed in a car accident on the road. A brief war between the USA and Spain led to the independence of Cuba and the loss of other Spanish territories in the Americas. The element neon was discovered at University College, London. The UK's 99-year lease of Hong Kong began, while the US annexed Hawaii. Births included English singer and actress Gracie Fields, German playwright Bertolt Brecht, Hungarian physicist Leó Szilárd, Swiss physicist Fritz Zwicky, Italian driver and car manufacturer Enzo Ferrari, Israeli prime minister Golda Meir, Dutch artist M.C. Escher, English sculptor Henry Moore, American composer George Gershwin, Belgian artist René Magritte and Northern Irish author C.S. Lewis. Among those who died were English writer Lewis Carroll, English artist Aubrey Beardsley, French painter Gustave Moreau, UK prime minister William Gladstone and German chancellor Otto von Bismarck.

Curie in a nutshell

Physicist and chemist

Legacy: radioactivity, medical use of both radioactivity and X-rays

Born Maria Salomea Skłodowska, 7 November 1867 in Warsaw, Poland

Educated: University of Paris

Married Pierre Curie, 1895

Nobel Prize in Physics, 1903

Marie Curie

First female professor at the Sorbonne, 1906

Founded the Institut du Radium (now the Institut Curie) in 1909

Nobel Prize in Chemistry, 1911

Died 4 July 1934 in Passy, France, aged 66

Element 96, curium, discovered 1944 and named after Curie and her husband in 1949

A strange energy

In 1895, German scientist Wilhelm Röntgen had been experimenting with cathode ray tubes – partly evacuated sealed glass tubes, which produced strange glows when an electrical current flowed through the tube between two internal metal plates. English physicist William Crookes, who did much of the early work with these tubes, had noticed that in some circumstances photographic plates kept near the tubes would be fogged as if they had been exposed to light, even though they were kept in opaque containers.

Röntgen discovered by accident that some form of ray appeared to be coming out of the tube at right angles to the flow of 'cathode rays' (what we now know to be a stream of electrons), emanating from the point where the rays were hitting a metal electrode. These rays had passed through the black cardboard that Röntgen was using to screen the side of his tube as if it were not there. The rays fogged a photographic plate Röntgen had stored apparently safely along-side the apparatus. Röntgen soon discovered the ability of these rays to pass through flesh, revealing the bones beneath as a shadow on a photographic plate. He referred to these mysterious new rays as *X-Strahlen* (X-rays), intended to be a temporary term, but one that stuck.

Next year, 1896, the French physicist Henri Becquerel made a similar accidental discovery. He had left a container of uranium salts sitting on a covered photographic plate. When the plate was later used, he found that the part of the plate that had been beneath the jar of salts was already blackened. It seemed that the compound was giving off a similar but stronger kind of emanation than an X-ray. However, unlike the cathode ray tube, the outcome was not dependent on putting electrical energy into the system. The uranium salts seemed to be capable of a spontaneous production of energy, at first sight breaking the first law of thermodynamics. This phenomenon, as we will see, would become known as radioactivity.

In the same year, shortly after, the English physicist J.J. Thomson, then head of the Cavendish Laboratory in Cambridge, had engaged his youthful New Zealander assistant Ernest Rutherford to investigate the nature of radioactivity. Rutherford excitedly wrote to his fiancée Mary, who was still in New Zealand: 'Don't be surprised if you see a cable some morning that yours truly has discovered a half a dozen new elements.' Following Becquerel's discovery, Rutherford would study the radioactivity of uranium.

At the time, radioactivity was thought to be a single emanation, but Rutherford was able to show that it had separate and distinct components. Some of the radiation from uranium was stopped by thin metal foil, while another part of it went straight through as if the foil were not there. In 1899, Rutherford called the less-penetrating radiation alpha rays and the more penetrating form beta rays. He was soon also to show that these 'rays' consisted of streams of electrically charged particles, as their paths could be deflected by electrical and magnetic fields.

Stranger in a strange land

The same year that Röntgen discovered X-rays, a young émigré in Paris, Maria Skłodowska, married her French fiancé, Pierre Curie. Known since then as Marie Curie, Skłodowska had studied at the Sorbonne and was working on magnetism. But by 1897, for her doctorate, Skłodowska (hereafter Curie) decided to examine the phenomenon that then was referred to as Becquerel rays or uranium rays. The aim of her doctorate was primarily to provide accurate measurements of the energy carried by these uranium rays. However,

Curie went further than was strictly necessary, testing other elements for emanations, including gold and copper.

After trying to detect rays from thirteen different elements there was still no result. The detection was performed by using a pair of metal plates, one of which was coated in a thin layer of the substance being studied. When an electrical voltage was applied across the plates, if the substance was giving off uranium rays, there would be an electrical current across the air gap between the plates, as the rays had the effect of ionising the air, stripping electrons from atoms making them electrically charged ions which can carry electricity. The size of the current gave an indication of the strength of the energy produced.

However, Curie had the inspiration of looking instead at the original source of the uranium, a black mineral called pitchblende. This substance had been mined for a number of years from Joachimsthal on the German–Czech border. Back in 1789, a German chemist called Martin Klaproth had extracted a greyish metal from pitchblende. This metal had proved useful as a yellow glass colouring and glaze constituent for pottery for centuries. It proved to be a new element, which Klaproth named uranium after the planet Uranus, discovered eight years previously by William Herschel.

Exactly why Curie tested pitchblende is not clear, though the expectation would have been that the level of emissions from pitchblende would be significantly lower than an equivalent amount of uranium, as there would be less of the ray-producing material present. However, the reverse proved to be the case. Pitchblende gave off around three times as much energy in the form of uranium rays as uranium itself. It was bizarre, as if diluting something made it stronger. Suspecting an error, Curie retested the pitchblende, and compared it with another mineral, aeschynite. Not only was the pitchblende definitely giving off more energy than its constituent, so did the aeschynite, which contains thorium but doesn't contain *any* uranium.

It seemed there was something other than uranium giving off these energetic rays in the pitchblende, and that thorium could also produce similar effects. The pitchblende did not contain a significant amount of thorium, but the ore was a complex mix of constituents that had never been fully analysed. It seemed there was something else even more powerful than uranium and thorium present. By now, Curie's husband Pierre was helping her after being turned down for a professorship at the Sorbonne. The Curies continued to try out different substances, hitting on a uranium-bearing mineral then known as chalcolite, though now more commonly called torbernite or copper uranite. This is primarily a copper/uranium phosphate and is found particularly in granite regions. Like the pitchblende, the chalcolite gave off around twice the energy in uranium rays as did pure uranium – but it was a significantly simpler mineral than pitchblende.

The Curies were able to produce a synthetic chalcolite, which proved to be less radioactive than the real thing, suggesting that it (and pitchblende) contained an unknown substance that was more energetic than uranium. Curie wrote up her findings in a note entitled 'Rayons Emis par les Composes de L'Uranium et du Thorium' (Rays Emitted by Compounds of Uranium and of Thorium), which was read at the French Academy of Sciences on 12 April 1898. Neither of the Curies was a member, which meant they were unable to speak themselves, but Marie's old professor, Gabriel Lippmann, was able to do so on her behalf.

In the note, Curie remarked that pitchblende and chalcolite were more active than uranium itself. 'This fact is most remarkable, and suggests that these minerals may contain an element much more active than uranium.' Curie concluded: 'To interpret the spontaneous radiation of uranium and thorium, one could imagine that all space is constantly traversed by rays analogous to Röntgen rays but much more penetrating and unable to be absorbed except by certain elements with high atomic weight such as uranium and thorium.'

The scene was set for the discovery that would win the Curies the Nobel Prize in Physics.

The naming of 'radio-activity'

The Academy showed relatively little interest in the potential for a new element, but Curie was certain there was the opportunity here to investigate something new. She also felt slighted that Becquerel, then the recognised expert on uranium rays, and who had helped them fund their researches, tended to ignore her and only dealt with Pierre. It's quite probable that this antagonistic environment was part of what drove Curie to persevere in what would prove to be a long, arduously painstaking piece of work.

With help from Pierre, she ground down 100 grams (3½ oz) of pitchblende and treated it chemically to try to separate off its different constituents. Each product would then be tested, and those with higher uranium ray energy were carried forward for further treatment. In two weeks, the Curies had a sample of what seemed to be the new active substance. However, the compound produced no unknown spectral lines (see box).

Spectroscopy

In the early part of the 19th century, a number of scientists had noticed that the spectrum of light from the Sun – the rainbow colours produced when sunlight is passed through a prism – contained dark lines, where a particular colour was missing. German physicist Joseph von Fraunhofer invented the spectroscope, a device to produce and magnify a spectrum so these lines could be studied. He discovered that other stars also had dark lines in their spectra, but they were not all in the same location as those produced by the Sun.

In the 1850s, German physicists Gustav Kirchhoff and Robert Bunsen discovered that bright-coloured spectral lines were produced in the glow when different elements where heated. (It was to produce an intense flame for analyses such as these that Bunsen's assistant Peter Desaga improved on Michael Faraday's design, marketing his product as the Bunsen burner, familiar from school laboratories.) Kirchhoff and Bunsen realised that these bright lines exactly corresponded to the positions of some of the dark lines in the Sun's spectrum. If an element gave off a particular colour when heated, then it would also absorb that colour when light passed through it when it was present in a star's atmosphere.

Spectroscopy would become – and still is, in more sophisticated forms – the standard mechanism for identifying elements. For example, the element helium was first discovered in the spectrum of sunlight, identified by English astronomer Norman Lockyer. If an unknown element was present in the sample extracted from pitchblende, Curie would have hoped to have seen new spectral lines emitted by it when it was heated.

Still convinced that there was something in pitchblende to discover, Curie asked for help from Gustave Bémont of the École Municipale de Physique et Chimie Industrielles, where Pierre had previously worked, and where the Curies were still provided with a lab. With his better equipment, Bémont was able to isolate a highly active substance from pitchblende. Marie and Pierre took over and, working in parallel, each seemed to have produced a more refined sample of the active substance – but each got a different figure for the current produced by their sample's ionising action. It was possible, then, that pitchblende contained not one, but two new elements.

Again, the Curies asked for assistance, this time from spectroscopy specialist Eugène Demarçay, but still it seemed to be the case that not enough of the substances had been produced, and there was no new spectral line seen. Even so, the Curies were positive about their findings, and by 13 July 1898, Pierre had written in his notebook that they believed they had found a new element, which he referred to as 'Po' – it would be named polonium after Marie's country of birth. Despite any issues Marie had with Becquerel's attitude, it was he this time who presented a paper to the Academy on their behalf.

The paper said that although they had not yet separated the active substance sufficiently to be able to detect its spectrum, it was 400 times as active as uranium. The Curies noted: 'We thus believe that the substance we have extracted from pitchblende contains a metal never before known, akin to bismuth in its analytic properties. If the existence of this metal is confirmed, we propose to call it polonium after the name of the country of origin of one of us.'

This paper was titled 'Sur une Nouvelle Substance Radio-active, Contenue dans la Pechblende' (On a New Radio-Active Substance Contained in Pitchblende). In writing this, Curie had given the phenomenon the name that it would continue to carry to this day (though it rapidly lost the hyphen). Radioactivity was soon to push 'uranium rays' and 'Becquerel rays' from the scientific lexicon. The 'radio' part of the name is not derived from the modern use of the term for a wireless receiver, but comes from the same source as radio: *radius*, the Latin for a ray.

Revealing radium: the 1898 day

It was usual for academics to leave Paris for a considerable time over the summer (the 'grandes vacances'), but by November, Curie had more clearly isolated the other radioactive substance in pitchblende, this being around 900 times as energetic as was uranium. Again, the Curies had assistance from Bémont with the chemistry and Demarçay for the spectroscopy – and, finally, the months of work resulted in new spectral lines. By December, Pierre had noted another new name in his notebook, a name derived from the term 'radio-active': radium.

Strictly speaking, at this point Curie was yet to isolate the element. The substance she produced was not pure, and the Curies were unable to distinguish the atomic weight of the substance, which was mostly barium, from pure barium metal. However, the decision was clear enough to enable publication of the landmark paper, co-authored with Bémont: 'Sur une Nouvelle Substance Fortement Radio-active, Contenue dans la Pechblende' (On a New, Strongly Radio-Active Substance Contained in Pitchblende). Once again it would be Becquerel who presented the paper on behalf of the Curies at the Academy, the day after Christmas Day, 1898.

The paper starts by mentioning polonium, but goes on to describe a 'second, strongly radioactive substance which is entirely different from the first one in its chemical properties'. Although the Curies had failed to isolate radium, they had good reason to consider that Marie had discovered a new element that was responsible for the radioactivity.

As they put it, 'M. Demarçay found a line in the spectrum which does not seem to belong to any other known element. This line, which is barely visible, using the chloride 60 times more active than uranium, becomes very prominent with the chloride which was enriched by fractionation up to the activity of 900 times uranium. The intensity of this spectral line thus increases at the same time as does the radioactivity, and this is, we think, a very important reason to attribute the radioactive emanation to our substance.'

Beyond the discovery of radium, the Curies hinted at the way that the behaviour of this substance seemed to break the first law of thermodynamics. They pointed out that the rays from the substance, like X-rays, caused the fluorescent substance barium platinocyanide to become luminous (though the effect was considerably weaker with the small amount of radium produced). The paper concluded: 'A source of light is thus achieved, and although, in truth, it is a very feeble light, it functions without any source of energy. This is a contradiction of the principle of Carnot, or at least appears to be.'

The source of radioactive energy

The Curies were making a remarkable claim that radioactivity appeared to contradict the 'principle of Carnot' which we would now call the first law of thermodynamics or conservation of energy, one of the fundamental laws of physics. As we saw above, Ernest Rutherford came up with the terms alpha and beta radiation, and in the early 20th century, working in Canada with English chemist Frederick Soddy, he developed the theory that radioactive decay was a result of the atoms releasing particles that resulted in a change in transmutation, transforming the element into a different one.

Rutherford moved on to Manchester University, where with Ernest Marsden and Hans Geiger he was able to show that atoms had a dense, positively charged nucleus. This structure provided the source for the alpha and beta particles. Meanwhile, Albert Einstein (see Day 6) would show that matter and energy were interchangeable. This made it possible to explain a source of the energy of radiation that avoided breaking the conservation law.

If matter could be turned into energy, then even a tiny reduction in the amount of matter in an atom could release a considerable amount of energy, apparently from nowhere. It was not, in fact, energy that was truly conserved, but the combination of matter and energy.

Leaving aside the slightly inaccurate attribution to their fellow countryman Carnot (as we saw in Day 3, Carnot's observations were only about heat and it was only later work by Clausius that moved the concept to cover energy) this paper flagged up the future direction for the Curies' work. Pierre focused on the physics behind this phenomenon, while Marie took on the task of producing pure samples of her discoveries, radium and polonium.

Up to this point, Curie had been working with laboratory-scale amounts of pitchblende, producing too small an amount of refined substance to truly isolate radium. Now she would work single-handedly on a near-industrial scale, processing around twenty kilograms (44 pounds) of pitchblende at a time, and in total dealing with several tonnes of the substance. The work was cold and hard in a huge old dissection laboratory that they had been assigned by the Sorbonne. Curie later said: 'the hangar was filled with great vessels full of precipitates and of liquids. It was exhausting work to move the containers about, to transfer the liquids and to stir for hours at a time, with an iron bar, the boiling material in the cast-iron basin.'

By 1902, though, Curie was able to announce that she had isolated a tenth of a gram of radium chloride.

Radium mania

A constant source of fascination to the Curies and others was that concentrated radium salts glowed in the dark. The Curies would visit the lab in the evening to see the ghostly blue glow, sent samples to other scientists, and kept a jar by their bedside. It was during this period that they and others began to notice the damage that radium could cause, burning the skin if kept close for too long.

Despite these early suggestions of danger, radium was treated as a wonder material, on the assumption that its spontaneous glow revealed its ability to donate healthy energy to those in contact with it. Radium was used in patent medicines. Natural hot water spas, which usually produce slightly radioactive water, were quickly relabelled radium spas. The British pharmacy chain Boots sold special soda syphon cartridges branded 'Spa Radium', which contained a very small quantity of radium, so that the gas they produced would irradiate the water as the fizz was added.

Such was the enthusiasm for radium that many products were claimed to contain it despite having no radioactive content whatsoever, simply to ride on the back of the commercial enthusiasm. Purchasers of these misleading products were arguably lucky. Others were unknowingly taking more of a risk. Radium salts, for example, were combined with fluorescent compounds and sewn into performers' costumes to provide a glow-in-the-dark spectacle which must have put those taking part in danger.

A US production featuring these radium-energised costumes was described as featuring 'fancy unison movements by eighty pretty but invisible girls, tripping noiselessly about in an absolutely darkened theatre and yet glowingly illuminated in spots by reason of the chemical mixture upon their costumes'. There is no doubt that the dancers were risking radiation poisoning, but the exposure of these performers to radiation was as nothing to those who produced luminous watch faces.

*Radium Radia, one of many medicinal cures
containing, or claiming to contain, radium.*

Women working in factories in the US producing glow-in-the-dark watch dials had been told that the radium-based radioactive paint they were using was harmless, and that they should bring their paintbrushes to a point to enable precision painting by using their lips or tongues. Many of the workers subsequently developed 'radium jaw' producing burns, bleeding and bone tumours.

Although knowledge of the dangers of radioactivity would eventually become commonplace, the Curies' paper did more than just reveal a new element, but rather opened up the study of radio-activity which would lead both to nuclear weapons and atomic power, along with a more complete understanding of the nature of atoms and their nuclei.

Marie Curie's death from aplastic anaemia at the age of 66 has often been ascribed to her exposure to radioactivity. While there is no doubt that her work with radioactive substances contributed to the risk of contracting this condition, it is now thought that a bigger contributory factor was her long-term work with X-rays at a time when the need for radiographers to use shielding was not well-understood.

Curie, the person

Maria Skłodowska was the youngest of five children, from a middle-class family in Warsaw. At the time, Warsaw was under Russian control and her parents' defiance of the regime seems to have come through in Curie's powerful urge to achieve. Her father, Władysław, was a science teacher and scientific matters fascinated Maria from an early age. At the time, Warsaw University did not take female students: she and her sister Bronia teamed up to apply for admittance to the Sorbonne in Paris.

Initially, Curie worked as a governess in Poland to support Bronia, until her sister had finished her studies. Curie then moved to Paris, changing her first name to the more French-friendly Marie. In the French capital, she lived with Bronia and her new brother-in-law for six months while she studied. At the time, Curie was one of only 23 women in the science faculty. She had intended to return to Poland as soon as she completed her degree, but she proved so successful that she was offered a scholarship to stay on at the Sorbonne.

It was in this period of her life, while searching for suitable laboratory space, which was at a premium, that she met Pierre Curie, who was by then already becoming known in his field. When Pierre proposed, Curie, who had been hurt in a previous relationship, felt it would be better to move back to Warsaw, but Pierre offered to leave France for her. This seems to have changed her viewpoint sufficiently to enable them to stay in Paris and marry.

Curie had two children, Irène (who would later also win the Nobel Prize for Chemistry) and Ève, who became a journalist and pianist. In 1906, Curie's family life was shattered when Pierre was killed in a street accident with a horse-drawn carriage. Curie was offered Pierre's professorship, becoming the first female professor at the Sorbonne.

In 1909, Curie started work on the Institut du Radium, which opened its doors in 1914. Funded by the Sorbonne and the Institut Pasteur, this had a pair of laboratories, dedicated to the study of radioactive elements and the medical applications of radioactivity. It was in a hospital here, established in 1922, that the therapeutic use of radioactivity in medicine was pioneered for the world. The organisation was renamed the Institut Curie in 1970.

Remarkably, after sharing a first Nobel Prize in Physics with Pierre and Henri Becquerel, Curie would win a second Nobel Prize, this time in Chemistry, awarded in 1911, 'in recognition of her services to the advancement of chemistry by the discovery of the elements radium and polonium, by the isolation of radium and the study of the nature and compounds of this remarkable element'. Curie was not only the first woman to win a Nobel Prize, she was the first person to win two prizes, and the only individual to win two different science prizes.

When the First World War broke out, Curie dedicated herself to the war effort. Initially her input was financial – she invested the prize money from her second Nobel Prize in French war bonds and tried to donate her medals to be melted down and sold, though the Bank of France turned this offer down. However, her biggest contribution was to set up mobile X-ray units to take the facility to troops in field hospitals. Initially her input was organisational, but by 1916 she had obtained a driving licence and started to drive mobile X-rays units and help hands-on herself. She got eighteen radiology cars into the field, which were used on over 10,000 soldiers, and set up a school to train female radiologists, which sent around 150 women out to support the medical work. By now, she was being aided by her daughter Irène, who first helped with organisation and training and by 1916 was also acting as a radiologist.

Without doubt, Curie was a remarkable woman who overcame the prejudices of the time to rise to the heights of a male-dominated scientific world. This comes across notably in the photographs from the famous Solvay conferences, which were attended by all the big names of the day in physics. The conferences were set up by Belgian industrialist Ernest Solvay, primarily as a vehicle to express his own, rather eccentric views. But the attendees were in the top ranks of physics and after politely listening to and ignoring Solvay they would turn to the major problems in their field.

Curie attended the Solvay conferences from the first in 1911, through to the fifth conference in 1927, which probably had the best-ever collection of eminent physicists in attendance. The photograph from the event shows well-known names including Albert Einstein, Erwin Schrödinger, Werner Heisenberg, Wolfgang Pauli, Lawrence Bragg, Paul Dirac, Louis de Broglie, Max Born, Niels Bohr and Max Planck. Amongst this sea of famous male faces sits a single woman. Marie Curie.

Life changers

Radiology (medical use of X-rays)

Although Curie was not involved in the development of X-rays, her championing of radiology during the First World War had a major impact on the spread of the use of X-rays for medical purposes.

Radiotherapy

Curie's discovery of polonium and radium was essential to the development of the medical uses of radioactivity in the treatment particularly of cancer. Radium in particular was the standard medical source other than the use of X-rays through to the 1950s. An equally important contribution was Curie's involvement in the Institut du Radium, now the Institut Curie, which pioneered many of the medical uses of radioactivity.

Tuesday, 21 November 1905

Albert Einstein – Publication of 'Does the Inertia of a Body depend on its Energy Content?'

It is probably no surprise that Albert Einstein appears here. But the paper published on this day was not the one that won him the Nobel Prize, nor his original paper on relativity – both of which were published earlier in the same year. It was instead a short piece that contemplated the impact of relativity on our understanding of energy and matter, which resulted in the equation $m = L/V^2$, soon to become a lot more familiar as $E = mc^2$. This paper, just three pages in length, contains the seeds of nuclear power – and the atomic bomb – applications that would go on to fascinate and terrify in equal measures. With an insight into this key period in Einstein's life at the time of his '*annus mirabilis*' (marvellous or miraculous year) of 1905, here is a day that transformed the world, if not always for the better.

The year 1905

To a scientist, this year stands out as Einstein's *annus mirabilis*, the year in which the then amateur theoretical physicist published four great papers, one of which, on the photoelectric effect, was foundational for quantum mechanics and would later win him the Nobel Prize. The rest of the world saw the opening of the Trans-Siberian railway, the first Russian revolution and the opening of the first Russian parliament, the foundation of Chelsea and Crystal Palace football clubs in London and the Automobile Association in the UK, the opening of the Simplon railway tunnel through the Alps, the foundation of Las Vegas, the establishment of Alberta and Saskatchewan in Canada, the independence of Norway from Sweden and the foundation of the Irish independence party Sinn Féin. 1905 saw the birth of English composer Michael Tippett, French fashion designer Christian Dior, Austrian celebrity Maria von Trapp, American author Ayn Rand, American actor Henry Fonda, French philosopher Jean-Paul Sartre, Swedish actress Greta Garbo, Italian composer Annunzio Mantovani, Belgian Queen Astrid and American millionaire Howard Hughes. Deaths included French writer Jules Verne and English actor Henry Irving.

Einstein in a nutshell

Physicist

Legacy: the special and general theories of relativity, quantum physics, $E = mc^2$, gravitational waves, lasers

Born 14 March 1879 in Ulm, Germany

Educated: ETH (Eidgenössische Technische Hochschule – Federal Institute of Technology), Zürich and University of Zürich

Began work at Swiss Patent Office, Bern, 1902

Albert Einstein

Married Mileva Marić, 1903

Published '*annus mirabilis*' papers including special theory of relativity, 1905

Published general theory of relativity, 1915

Married Elsa Löwenthal, 1919

Nobel Prize in Physics, 1921

Emigrated to USA and took up position at Institute for Advanced Study, Princeton, 1933

Died 18 April 1955 in Princeton, New Jersey, USA, aged 76

Element 99, einsteinium, discovered 1952 and named after Einstein in 1955

....

The disappearing ether and contractions

As we saw in Day 4, James Clerk Maxwell never lost his enthusiasm for the ether as the medium through which light travelled. He would write, 'Whatever the difficulties we may have in forming a consistent idea of the constitution of the ether, there can be no doubt that the interplanetary and interstellar spaces are not empty, but are occupied by a material substance or body, which is certainly the largest and probably the most uniform body of which we have knowledge.' However, Maxwell's explanation of electromagnetic waves had fatally damaged the theoretical requirement for the ether. This step forward on the theory side would be supported, unintentionally, by experimental evidence provided by American physicists Albert Michelson and Edward Morley.

The pair were not setting out to disprove the existence of the ether, but rather to show that it really did exist, despite doubts. As we have seen, the ether was thought to be a universal medium, filling all of space, which allowed the waves of light to cross the otherwise empty vacuum. If the ether existed, then the Earth was moving through it. And if that were the case, Michelson and Morley believed that it should be possible to detect a difference in the speed of light measured on the Earth, depending on the direction in which the measurement was made, as the Earth's movement through the ether should add to light's speed.

The experiment, set up in 1887 at what is now Case Western Reserve University in Cleveland, USA, was made up of a device called an interferometer fixed onto a slab of stone over a metre across. This slab sat on a wooden circle, which floated in a trough of mercury fixed to a brick base. The idea was to provide a vibration-free piece of equipment which could be set up to turn extremely slowly, taking a whole six minutes for a single rotation.

The interferometer first split a beam of light, then sent each half back and forth between sets of mirrors, the two beams travelling at right angles to each other. The beams were then recombined. When light beams come together, depending on the position of each wave in its cycle, the two beams will interfere, either reinforcing each other or cancelling out. As a result, the combined beam produced a series of light and dark fringes that were observed through a microscope.

If, as was assumed, the Earth's movement through the ether on its orbit, travelling at around 30 kilometres (18.6 miles) per second, changed the speed of light slightly, it would be expected that there would be a small difference in the time taken for light to traverse the two arms of the interferometer. As the stone slab rotated, the relative speeds on the two routes would change, depending on the light beams' alignment with the Earth's direction of movement. As a result, the interference fringes should shift over time in a cyclic fashion.

Interfering with gravity

The interferometer used by Michelson and Morley would provide the model for observatories for the remarkable phenomenon of gravitational waves. These vibrations in space and time, predicted by Einstein in 1916, are sent across the universe by interactions between massive bodies, such as the collision of black holes.

Despite decades of attempts to detect gravitational waves, scientists failed to do so until the establishment of the LIGO observatory, which made its first successful detection in 2015. LIGO uses two vast interferometers several kilometres long, and located thousands of kilometres apart in America. When gravitational waves hit such an interferometer, they very slightly change its length, resulting in the kind of shift in fringes that the Michelson–Morley experiment hoped to detect.

The changes in length of the LIGO device involved are tiny – smaller than an atom – but it and other such observatories have now detected many gravitational wave events.

Contrary to their expectations, Michelson and Morley observed no change. The rotating device was not causing the fringes to shift. This was problematic. Science is usually said to work more by inference and falsification than positive proof. We observe what has happened, make assumptions about what will happen in the future and test those assumptions. If they fail, we are able to disprove a theory; if the assumptions repeatedly succeed in prediction, by induction we assume that our theory is correct, until and unless other evidence becomes available.

Unfortunately, in a case like this, falsification is much harder because the effect the experimenters expected to see was on the borders of what could be detected with the equipment of the day. Unless the experimenters had been very careful it would have been entirely possible that the effect was there, but not detected (just as the spectro-scopes initially failed to detect new lines in Marie Curie's experiments in Day 5). However, that was the point of building such a massive, stable piece of apparatus. It should have been sufficiently sensitive to produce a result. In practice, only a tiny change was detected, and this was too small to either support the speed of the Earth through the ether or to exclude the error levels of the apparatus. A number of other experiments were made over the following years, but still to no avail. There was no evidence that the ether existed.

At least, this was one interpretation of the results. However, like Maxwell, many physicists were reluctant to give up on the ether. In 1889, the Irish physicist George FitzGerald came up with an ingeni-ous idea to explain why Michelson and Morley had failed to find any shift. FitzGerald suggested that moving objects became shorter in their direction of motion, based on the relationship between elec-tromagnetic forces and movement. Three years later, Dutch physicist Hendrik Lorentz independently came up with a similar concept.

The combined theory became known (somewhat unfairly, given that FitzGerald got there first) as Lorentz–FitzGerald contrac-tion. By 1904, Lorentz had expanded the theory, but its basis on movement through a stationary ether was about to be blown apart by Albert Einstein.

Thinking in a relative way

We tend to think of Einstein as an ageing man with a shock of white hair, recognised throughout the world as the epitome of scientific genius. This was anything but the Einstein who took FitzGerald and Lorentz's ideas to the next level and disposed of the ether. In 1905, Einstein was 26, and was yet to achieve an academic post. He was working as a clerk (third class) in the Swiss Patent Office in Bern.

When Einstein's paper on the special theory of relativity, 'Zur Elektrodynamik bewegter Körper' (On the Electrodynamics of Moving Bodies), was published on 26 September 1905, it came up with the same contraction result as Lorentz's – but as a small part of a more radical shift in understanding of the nature of space and time. For Lorentz, the ether provided an unmovable spatial reference frame. Many years before, Isaac Newton had argued for 'absolute' time and space – a fixed background against which everything occurred, and which underpinned Lorentz's concept of a fixed ether. But Einstein threw out this concept. In the special theory, there was no fixed thing against which everything could be measured. All positions and movement were relative. Measurements in space and time could be based on any viewpoint, apparently static or moving. Each viewpoint (or 'reference frame' in physicists' terms) had equal validity.

All that Einstein needed to achieve this new way of looking at things was the combination of Maxwell's realisation that light would always travel at the same speed in a particular medium and traditional Newtonian mechanics. This was a major breakthrough, although it was a theory that was in the air. It really only took the rejection of the ether to make it possible to move on from Lorentz–FitzGerald contraction to a more comprehensive description of the impact of light's behaviour on relativity.

The special theory of relativity predicted a range of effects when objects are in so-called inertial frames – situations where the objects are not under the influence of acceleration. As well as contraction in the direction of motion, the theory predicted that objects would also increase in mass and have a slower passage of time (a concept known as time dilation). However, the nature of relativity doesn't fit well with the English language. It's tempting, for example, to say that someone in motion *experiences* a slower passage of time. But they don't. From the viewpoint of the 'moving' person, they are not in motion. They are still, and the universe around them is moving in the opposite direction.

Because no frame of reference is privileged to especially denote being stationary, what we are saying is not that the moving person experiences length contraction, mass increase and time dilation. Rather, when observed to be moving by someone else, from the viewpoint of that observer the moving person undergoes length contraction, mass increase and time dilation. Note that this does not mean that the moving person only *appears* to undergo these changes. They really happen, from the point of view of the observer.

True time machines

The effect of time dilation makes it possible to produce a working time machine, if not one that behaves like the classic device in science fiction. If a spaceship leaves Earth at high speed, because time is running slower on the spaceship than it is on Earth (from the viewpoint of Earth), after the ship has been travelling for some time, then anyone on the ship will have aged less than those who are left behind.

Initially, this effect is symmetrical. As far as the people on the spaceship are concerned, it is time on Earth that is running slowly, as in the travellers' frame of reference, the spaceship is not moving. However, to be able to make a round trip, the ship undergoes acceleration that the Earth doesn't, in order to change direction. This effectively resets the clocks. As a result, on returning to Earth the space travellers will really have travelled into their future.

We don't usually see such effects, as it's necessary to travel very quickly to have a significant time dilation effect. Our best time machine to date is Voyager 1, which has travelled around 1.1 seconds into the future. But to achieve significantly more of an effect would require speeds of more than 10 per cent of the speed of light, which are yet to be practically possible.

The special theory of relativity was remarkable stuff, but had a mixed reception at the time. Einstein would win the Nobel Prize not for special relativity or for his far more sophisticated general theory of relativity, which brought acceleration and gravity into the mix, but for his explanation of the photoelectric effect. Yet over time, the effects of the special theory would be repeatedly proved by experiment and those who clung on to the idea of the ether were left behind.

Although the special theory is important as a closer description of reality than Newtonian mechanics, it only departs significantly from what Newton had predicted at high speeds. Of itself, it had relatively little impact on everyday life. But Einstein had not finished with the impact of special relativity. A couple of months later he published a very short paper that was effectively an addendum to the theory. And this would go on to have dramatic consequences.

A powerful addendum: the 1905 day

The paper, entitled 'Ist die Trägheit eines Körpers von seinem Energieinhalt abhängig?' (Does the Inertia of a Body Depend on its Energy Content?) was received by the journal *Annalen der Physik* on 27 September 1905 and published on 21 November of that year, making it the last of Einstein's *annus mirabilis* papers. It's important to dig into the detail of this paper, both to understand how Einstein achieved his result and to appreciate the relative simplicity, but striking impact of this landmark publication.

The paper begins by making it clear that we are moving on from his special relativity paper 'Zur Elektrodynamik bewegter Körper' (On the Electrodynamics of Moving Bodies) by saying, 'The results of an electrodynamic investigation published by me recently in this journal lead to a very interesting conclusion, which will be derived here.'

Einstein then gives us his starting points. In the earlier paper, he had made use of the constant velocity of light and the principle that physical laws don't depend on which of two bodies moving steadily with respect to each other is being referred to. Based on these principles, he had derived an equation that showed how the energy of a beam of light changes if you move with respect to it.

This is a key point. Usually, when we move with respect to something else that is moving, the relative speed of that object changes from our viewpoint. So, for example, if I drive towards a car at 50 miles per hour, and that car is also driving towards me at 50 miles per hour, the relative speed at which the cars come together (from either driver's viewpoint) is 100 miles per hour. However, if I drive towards or away from a beam of light it still comes at me at exactly the same speed as if I were stationary. This doesn't mean, though, that there is no change at all.

Think for a moment of that light as a wave. If I am moving towards it, I get closer to the source between each peak in the waves of the light, so the wave gets squashed up, giving it a shorter wavelength from my viewpoint. The light undergoes a shift towards the blue end of the spectrum. Similarly, if I move away from the light, the wavelength is stretched, so it has a red shift. The shorter the wavelength, the greater the energy a photon of light carries. So, when I move towards a beam of light, it doesn't travel any faster, but its energy increases.

Einstein shows the change in energy in his paper using this equation:

$$l^* = l \frac{1 - \dfrac{v}{V} \cos \varphi}{\sqrt{1 - \left[\dfrac{v}{V} \right]^2}}$$

It looks a bit messy, but what's going into this equation is quite simple (even if Einstein's choice of symbols is not ideal from a modern viewpoint). We can ignore the 'cos φ' part. This is just allowing for the possibility that the light isn't coming straight at us – φ here is the angle the light ray is pointing away from heading straight on. Without that, we have a straightforward relationship between the light's energy with and without our moving (l^* and l respectively) and the speed we are moving at (v) divided by the speed of light (V).

Anyone who has ever looked into special relativity will recognise in that square root divisor an amount usually given its own symbol, γ, as it turns up so frequently in special relativity calculations. It's the factor that enables us to calculate the impact of time dilation, length contraction and mass increase.

Einstein then imagines a particular setup where a body loses some energy by emitting two beams of light, which head off in opposite directions. He calculates the energy of the body as seen from its own frame of reference (where the body is at rest), and from a moving frame of reference, where the body has a velocity. He then does the same for the energies of these bodies after emitting the light, which will be reduced by the amount of energy lost in the form of light. These will differ because of the change in energy of the light in the equation above. From this, Einstein is able to deduce the change in kinetic energy due to emitting the light, which he makes:

$$\frac{L}{V^2} \frac{v^2}{2}$$

Here, L is the energy of the light, v is the speed of movement and V is the speed of light. The kinetic energy of a moving body, as we are taught at school, is $\frac{1}{2}mv^2$. So, if this is the equivalent of the equation above, we can see that we get a reduction of mass of L/V^2 when the light is emitted. Einstein is telling us that $m = L/V^2$.

Let's put that in more familiar symbols. He is saying that $m = E/c^2$. Or to rearrange that a little, $E = mc^2$.

Finally, Einstein makes it clear that this is not just about the emission of light. 'Since obviously here it is inessential that the energy withdrawn from the body happens to turn into energy of radiation rather than into some other kind of energy, we are led to the more general conclusion: The mass of a body is a measure of its energy content.' He goes on to say that if the energy changes by a certain amount, the mass changes by that amount divided by the square of the speed of light in appropriate units.

Finally, Einstein gives us an observation based on the relatively recent discoveries of Marie Curie on radioactivity that we encountered in Day 5. 'Perhaps it will prove possible to test this theory using bodies whose energy content is variable to a high degree (e.g., salts of radium).'

Now I am become Death, the destroyer of worlds

Robert Oppenheimer, the head of the Los Alamos laboratory where the atomic bomb was developed during the Second World War, said that the line 'Now I am become Death, the destroyer of worlds' from the Hindu scripture the *Bhagavad Gita* came to his mind when he witnessed the first test detonation of a nuclear weapon.

That weapon has a direct link to Einstein's three-page paper. In a nuclear fission reaction, an atomic nucleus splits, resulting in a reduced overall mass of matter, which is emitted in the form of released energy. Of itself, this wasn't enough to make a nuclear weapon. Because of the 'c^2' part of the equation (the speed of light is a big number), the energy released is large even for a small amount of mass lost. But the mass of a single atomic nucleus is tiny – and the overall reduction in mass is ludicrously small. For example, when a nucleus of uranium-235* decays, it produces around 200 MeV (mega electron volts) of energy, in the energy unit used by particle physicists. This is around 30 trillionths of a joule. To put that into context, a typical LED light bulb puts out around 4,000,000,000 times as much energy every second.

What was required to take Einstein's revelation to a deadly practical level was the concept of the chain reaction, dreamed up, according to its originator, Hungarian physicist Leó Szilárd, while he waited for a change of traffic lights. Szilárd was staying in the Imperial Hotel in Russell Square, London at the time and was waiting to cross the road where Southampton Row enters the square. Running through his mind as he waited was a dismissive remark by Ernest Rutherford.

In 1933, when interviewed by the US newspaper the *Herald Tribune*, Rutherford had remarked, 'The energy produced by the breaking down of the atom is a very poor kind of thing. Anyone who expects a source of power from the transformation of these atoms is talking moonshine.' This makes sense, given the minuscule amount of energy produced by a decaying atom. 'But,' Szilárd said, reflecting on his inspiration, 'it suddenly occurred to me that if we could find an element which is split by neutrons, and which would emit two neutrons when it absorbed one neutron, such an element, if assembled in sufficiently large mass, could sustain a nuclear chain

---

* This is an isotope of uranium with a total of 235 protons and neutrons in the nucleus. Isotopes of an element all have the same number of protons, but different numbers of neutrons. Uranium-235 is the isotope required for a nuclear chain reaction.

reaction.' It was like getting compound interest on a bank account. You invested one neutron, and got two neutrons in return. Now both of these were invested and you got four – and so on. The process could be self-sustaining in its generation of energy if kept in check, or it could run away, doubling in its rate with every reaction.

The controlled, self-sustaining chain reaction is the basis of nuclear energy which, despite the understandable fears of nuclear accidents, is a green source of energy that has killed far fewer people than fossil fuels such as coal have for the same amount of energy produced. The runaway reaction was the one that Oppenheimer and his teams would release in the Trinity atomic bomb test and in the two bombs dropped on the Japanese cities of Hiroshima and Nagasaki.

Although Einstein was persuaded to write to American president Roosevelt affirming the necessity to undertake the nuclear research leading to the development of the atomic bomb because of fears that developments were already underway to produce such a device in Nazi Germany, he would always regret the translation of his simple observation into such a deadly weapon.

Einstein, the person

Born into a happy middle-class family, Einstein was an independent thinker from an early age, an approach that would never leave him. Throughout his education he received very mixed reports. If a subject interested him, he would dedicate huge effort to it – if it didn't, he would go out of his way to avoid involvement. He was far happier doing things his own way than following the crowd and sticking to the rules – an uncomfortable viewpoint for someone living in the regimented Germany of the last years of the 19th century. Secondary school was dull; Einstein was told that he was a lazy boy who would amount to nothing. Things came to a head, though, with a family move when Einstein was fifteen.

Einstein's father, Hermann, was struggling to keep his business afloat. Hermann's more successful brother Jakob suggested that the family should move to Italy and set up business there, which they did, leaving Einstein in a boarding house in Munich to continue his schooling. Within six months, Einstein was sufficiently fed up that he obtained letters from his doctor and his maths teacher saying that he was not getting anywhere with his schooling and was in danger of mental breakdown. Einstein took these to the headmaster, saying he was leaving: the head retorted that Einstein was expelled anyway.

Heading to Pavia in Italy to join his family, Einstein appeared to be in good spirits, but perhaps made it clear why he had made this move when he renounced his German citizenship a number of months before he would have been obliged to spend a year undertaking military service. Einstein now needed somewhere to go. He applied to and was rejected by the prestigious Zürich university the Eidgenössische Technische Hochschule (ETH) – the Swiss Federal Institute of Technology. He had attempted to get in a year before the usual age of application, so spent a year in a Swiss school and on his second attempt scraped through the ETH entrance exams. As before, his maths and sciences were at a high level: it was the humanities subjects than let him down.

After spending far too little time on his studies at the ETH, Einstein was awarded an underwhelming degree – but perhaps more important to him at the time, he met Mileva Marić, a fellow student hailing from what is now Serbia. They became a couple and in 1902 Marić gave birth to a daughter, Lieserl. Unmarried, and with limited income from the temporary job as a teacher that Einstein had taken, the couple appear to have put Lieserl up for adoption – her existence was hushed up and not publicly revealed until the 1980s.

In the summer of 1902, Einstein got the job at the patent office where he would have his remarkable year, giving him a stable job so that he and Marić were able to marry in January 1903. He remained at the office until 1909, by which time he had sufficient academic status to take on a string of university posts. Most of the time, Einstein was living separately from his family, who stayed behind in Switzerland (he had two sons with Marić, Hans Albert born in 1904 and Eduard born in 1910). In 1917, exhausted by his work on the general theory of relativity while at Berlin University, and with the establishment of the Kaiser Wilhelm Institute for Physics of which he was the director, Einstein was housebound for almost a year, looked after by his cousin Elsa Löwenthal (an Einstein herself before her first marriage, from which she was by this time divorced).

Affection grew between the cousins and Einstein negotiated a divorce with Marić, agreeing to the condition that she would receive the money should Einstein win the Nobel Prize. He married Löwenthal in June 1919. After a comfortable time in the 1920s, the rise of Hitler gave Einstein, from a Jewish family though not a practising member of the religion, a new and pressing reason to consider moving on. In October 1933, the Einsteins moved to Princeton, New Jersey in America, where he would live and where he would work at the newly founded Institute for Advanced Study for the rest of his life.

In his years in America, Einstein did not make any scientific breakthroughs, but helped a number of younger physicists in their careers. He was asked to become the first president of the newly founded state of Israel, but turned the position down. Although he was political in the sense of being, for example, a vocal pacifist, he did not want to dedicate his time to a political life. One aspect of Einstein's early years that continued was his interest in music. He was a decent violinist and frequently took the opportunity to play.

Life changers

Nuclear power

The positive outcome of Einstein's paper on the relationship between mass and energy was the development of nuclear power. Although the industry has a mixed reputation due to a handful of accidents, it has generally been very safe and has killed far fewer people than fossil fuels have for the equivalent energy production. Nuclear energy continues to have a future as we move to low-carbon fuels to mitigate the impact of climate change.

Atomic weapons

It isn't possible to discuss this paper of Einstein's without the spectre of atomic weapons arising. The first atomic weapons were fission weapons, using exactly the principle Einstein discussed, combined with the concept of the chain reaction. Since the 1960s, however, the majority of nuclear weapons have been fusion bombs (so-called hydrogen bombs), which do make use of mass–energy equivalence, and usually have a fission bomb as a trigger, but work on the basis of nuclear fusion, the energy source of the Sun, rather than fission for their main detonation.

Saturday, 8 April 1911

Heike Kamerlingh Onnes – Discovery of superconductivity

Often, it is hard to pin down a specific date for a discovery – but in the case of Dutch physicist Heike Kamerlingh Onnes, we can fix his discovery of superconductivity from his lab notebooks, scrawled in his appalling script. Superconductivity, where electrical resistance disappears at extremely low temperatures, makes it possible to construct ultra-powerful magnets and transmit electrical currents without loss, finding applications in facilities such as the Large Hadron Collider, hospital MRI scanners and ultra-high-speed levitating trains. The discoveries arose arguably despite rather than because of Kamerlingh Onnes' personality and management style. His approach was outdated, considered even by the standards of the time to be paternalistic and overbearing. Yet his discoveries would have long-term impact – and promise more to come.

The year 1911

This year saw the first official airmail flight, the maiden voyage of the *Titanic*'s sister ship RMS *Olympic*, the coronation of British

King George V, the founding of the Mars confectionery company in Tacoma, Washington, the rediscovery of Machu Picchu in Peru by Hiram Bingham, the theft of the Mona Lisa from the Louvre, the Wuchang uprising that began the revolution leading to the founding of the Republic of China, the launch of car maker Chevrolet, Ernest Rutherford and colleagues discovering the atomic nucleus and Roald Amundsen reaching the South Pole. Born this year were American actor Danny Kaye, American actor and president Ronald Reagan, American playwright Tennessee Williams, American actor Vincent Price, French president Georges Pompidou, American physicist John Wheeler, American actress Ginger Rogers, Canadian author Marshall McLuhan, American actress Lucille Ball and English writer William Golding. Deaths in this year included English scientist Francis Galton, Austrian composer Gustav Mahler and English dramatist W.S. Gilbert.

Kamerlingh Onnes in a nutshell

Physicist

Legacy: superconductivity

Born 21 September 1853 in Groningen, Netherlands

Educated: Heidelberg University and University of Groningen

Professor of Experimental Physics at the University of Leiden, 1882–1923

Married Maria Bijleveld, 1887

Founded what is now the Kamerlingh Onnes Laboratory, 1904

Nobel Prize in Physics, 1913

Died 21 February 1926 in Leiden, Netherlands, aged 72

Heike Kamerlingh Onnes

The scales of temperature

As we saw with the caloric theory and the development of thermodynamics in Day 3, heat and temperature are topics that were experienced before they were understood. The first real thermometers were only introduced in the 18th century, which is also when the most familiar temperature scales, Fahrenheit and Celsius were introduced. It's strange to think that people we now regard as scientists, such as Galileo or Isaac Newton, had no modern conception of temperature. The Fahrenheit and Celsius scales are practically useful, but have an inbuilt flaw.

The fact that the temperature scales we are most familiar with are deceptive can be shown in a misunderstanding that happens regularly in news reports. Let's take something from the field of climate change – average global temperature. This is the measure of global warming. In 1900, this temperature was around 13.8°C. It has since increased by about 1.1 or 1.2°C. If left unchecked, it's possible that the rise could be as much as 5 degrees by the end of the 21st century. Such a rise is sometimes portrayed as a 36 per cent rise, because 5 is 36 per cent of 13.8.

Usually in this book, temperatures have been given in both Celsius and Fahrenheit. The Fahrenheit equivalents were intentionally omitted in the previous paragraph because they demonstrate the problem. Using Fahrenheit, the average temperature in 1900 was 56.8°F, and the potential increase by the end of current century is 9 degrees. Undertaking the same calculation, the percentage rise would be 16 per cent. Somehow, changing temperature scales appears to have halved the impact of climate change – which seems, at the very least, unlikely.

The reason this confusion occurs can be highlighted by comparing objects in a freezer and a fridge. My freezer keeps things at

−18°C and my fridge at 4°C. What is the percentage difference in temperature? The way we usually work out a percentage increase is to take the increase, here 22, multiply by 100 (hence 2,200) and divide by the starting value. Here that number is −18, producing a percentage increase of −122.222... which makes no sense. It is impossible to specify a percentage change on a scale that doesn't start at zero. In the real world, just as you cannot possess a negative number of objects, nor can there be a negative temperature.

The end of coldness

If there were a temperature scale that started at zero, there would have to be a lower limit of coldness – and surprisingly, the idea that such a state was possible cropped up before modern thermometers, temperature scales, or a useful definition of what temperature was, other than an indication of how cold it felt. Newton's contemporary and early chemist Robert Boyle spent a chapter of his magnificently titled 1665 book *New experiments and observations touching cold, or, An experimental history of cold begun to which are added an examen of antiperistasis and an examen of Mr. Hobs's doctrine about cold* addressing the concept of the '*primum frigidum*'.

Admittedly, Boyle's main aim was to counter the possibility of there being 'some Body or other, that is of its own nature supremely Cold, and by participation of which, all other cold Bodies obtain that quality'. He concluded that 'my design in playing the Sceptick on this subject, is not so much to reject other mens probable opinions, of a primum frigidum, as absolutely false, as 'tis to give an account, why I look upon them, as doubtful.' However, the fact was that Boyle felt it sensible to spend 52 pages questioning the concept.

In effect, the *primum frigidum* was conceived as a kind of anti-caloric substance. As such, Boyle's doubts were realistic. However, when thermometers came in during the early 1700s, the lower limits of temperature became an area of more quantitative speculation. The French natural philosopher Guillaume Amontons constructed a thermometer that measured temperature as the height to which a quantity of air would hold up a column of mercury. As the temperature fell, the volume of air decreased. Clearly there had to be a limit, as that volume could not fall forever. This fitted well with the theory of heat of the time. If heat reflected the amount of caloric fluid in a body, there must be some point where, having lost all its caloric, there was nowhere left to go. As made clear in a 1798 note from the Manchester Literary and Philosophical Society, the first use of the term 'absolute zero' reflected the idea of there being a 'point of absolute privation of caloric'.

There were several attempts to calculate this limit – some in the region of –270 to –240°C (–454 to –400°F) ranging all the way down to John Dalton's estimate of –3,000°C (–5,368°F). However, the realisation that made it possible to more accurately pin down this lower limit (and to establish beyond doubt its existence) came through the development of thermodynamics. With the understanding that temperature is a measure of the kinetic energy of the motion of atoms and molecules (later qualified to include the energy levels of electrons around atoms), there was a clear minimum of absolute zero, where all such energy is at a minimum.

The Scottish physicist William Thomson (Lord Kelvin) reflected this understanding in the development of a new temperature scale, the absolute or Kelvin scale, which starts at 0 for absolute zero (–273.15°C or –459.67°F) and goes upwards in units the same size as degrees Celsius. The units of this scale are now defined as kelvins. So, the freezing point of water, for example, is 273.15 K. (Note that the units are kelvins, not *degrees* Kelvin.)

How low can you go?

If there were to be such a lower limit of temperature, there was more than enough reason to attempt an investigation of this extreme. The challenge was how to drive temperatures down below the ambient temperature. In 1758, a collaboration between the English chemist John Hadley and American Benjamin Franklin at the University of Cambridge produced the first artificially generated temperatures well below the freezing point of water by using evaporation of volatile liquids such as ether and alcohol. Evaporation reduces temperature because energy is needed to complete the transition from liquid to gas – it's why fans cool us by evaporating sweat from the skin. Hadley and Franklin got their experiment down to around –14°C (6°F).

One of Michael Faraday's many achievements was to liquify a range of gases, using a combination of high pressure and low temperature, though he thought some gases, such as oxygen and hydrogen, could not be made liquid. This process continued as scientists used existing lowest temperatures as starting points to produce even lower ones, leading to the triumphant work of Heike Kamerlingh Onnes. In July 1908, in his Leiden laboratory, he managed to liquify the most resistant gas, helium, reaching a temperature of around 1.5 K (–271.65°C or –456.97°F).

It was for this that Kamerlingh Onnes would win the Nobel Prize in Physics for 1913, the record stating that it was 'for his investigations on the properties of matter at low temperatures which led, inter alia, to the production of liquid helium'.

We have since gone further, reaching incredibly close to absolute zero – around a ten-billionth of a kelvin – but it should be stressed that the final limit will never be achieved. The third law of thermodynamics, which is technically about changes in entropy, means that there is no way to reach absolute zero in a finite number of steps.

There are occasionally experiments undertaken that appear to produce temperatures below absolute zero, but this is not the case in any meaningful way.

What these 'negative absolute temperatures' reflect is the use of entropy in thermodynamics. The circumstances in which such an effect occurs are having a gas in which most of the particles have very high energy (though not kinetic energy). The combination of high energy and low number of ways to organise the constituent parts means that the usual distribution of energies in a gas is inverted; this has been, somewhat artificially, represented as having a negative absolute temperature, despite the contents not being cooled below absolute zero.

The transition: the 1911 day

Having successfully liquified helium, Kamerlingh Onnes was determined to discover the impact of such drastically low temperatures on the behaviour of materials. Measuring how well solid mercury conducted electricity at low temperatures, he discovered a shocking transition occurred at 4.2 K. When a substance conducts electricity, some of the electrical energy is lost as heat to resistance – which we now know to be an interaction between the electrons carrying the electrical energy and the structure of the material. But at this temperature, Kamerlingh Onnes discovered that mercury suddenly stopped having any electrical resistance whatsoever.

For most of the key dates in this book we are reliant on the publication date of a paper announcing the discovery to the world. However, in the case of superconductivity, we are not making do with the date of publication of his paper 'On the Sudden Rate at Which the Resistance of Mercury Disappears', which appeared in 1912. This is because Kamerlingh Onnes put some very precise timing in his laboratory notebook. His note-keeping does not make uncovering this information easy, though. Not only was the record of his 1911 discovery in a notebook labelled 1909–1910, the content is scrawled in appallingly bad handwriting, written in pencil.

Timing his entry at exactly 4pm on 8 April, Kamerlingh Onnes notes '*Kwik nagenoeg nul*'. This doesn't initially appear to be a highly meaningful observation, roughly translating as 'Quick pretty much null'. But 'quick' here is a shortening of 'quicksilver' – mercury – and he is saying there was no detectable resistance. It's from this entry that we get 8 April, but there is no year attached. Worse still, somewhat later in the book we get a note of an experiment that he dated 19 May 1910. Given the labelling of the notebook, this would seem to suggest that the 1911 date is incorrect, but in reality, he did not have the equipment to undertake the experiment until April 1911.

It took a while for theory to catch up with experiment. At the time it was already known that a flow of electrons through a conductor carried electricity, behaving something like a gas under pressure. As gases get colder, their components get less mobile. So, the assumption of many at the time was that electrons would grind to a halt as absolute zero was approached, making resistance shoot up. Kamerlingh Onnes, though, was in a relative minority expecting resistance to fall, but only believing that were would be zero resistance at the (unachievable) absolute zero. Kamerlingh Onnes would mostly refer to the phenomenon as 'supraconductivity', though occasionally he used 'superconductivity' – both terms continued to be employed for some decades, though superconductivity has now become standard.

The implication of zero resistance is, for example, that an electrical current flowing around a loop of such material would continue forever without any additional energy being put into the system. Bearing in mind the difficulty of achieving such a low temperature, it was not trivial to demonstrate that electrical resistance had become exactly zero. A typical way of measuring the resistance of an object is to put a known voltage across it and measure the current, but here, a meter would simply go off the scale. From school, you may remember the relationship between electrical voltage, current and resistance, is supplied by the simple equation $V=IR$.

This means that when you put a voltage of V across a resistance of R you get a current of V/R. However, if R becomes zero, the current attempts to become infinite. Something has to give. In practice, it would be discovered that such currents are self-limiting. Not only is there a finite supply of electrons to carry the current, the build-up of magnetic field will eventually disable the superconductivity; however, it doesn't stop the current shooting up to the extent that metering it was impractical.

There are two significant practical implications of this disappearance of resistance. One is that without resistance, there is no loss to heat. At the moment, all electrical power lines lose energy to heat, but were it possible to make a power line superconducting, it would transmit all the energy. Secondly, the strength of an electromagnet is dependent on the current that flows through it. Electromagnets constructed from superconducting materials can produce a far stronger magnetic field than conventional magnets, which has proved useful in everything from magnetic resonance imaging scanners to particle accelerators and magnetic levitation trains.

The Meissner effect

One interesting feature of superconductors is the Meissner effect, discovered in 1933 by German physicists Walther Meissner and Robert Ochsenfeld at the German National Institute of Natural and Engineering Sciences, the Physikalisch-Technische Bundesanstalt in Berlin.

As we saw in Day 2, Michael Faraday came up with the concept of electrical and magnetic fields. Generally speaking, a magnetic field fills space and passes through objects, even though it can be distorted by electromagnetic effects. But at the transition temperature where superconductivity kicks in, a conductor will suddenly entirely expel the magnetic field within it, forcing it outside the material. This provides one of the more dramatic lab demonstrations of the effects of superconductivity. An ordinary permanent magnet sitting on top of an electrical conductor will start to levitate above the conductor if the conductor becomes superconducting and expels the magnet's field.

Rather than attempt to measure resistance using a meter, Kamerlingh Onnes set an electrical current in motion around a loop of semiconductor and measured the magnetic field produced by this very basic electromagnet. If the loop had any resistance, then the current would gradually drop as heat was produced, reducing the magnetic field. Kamerlingh Onnes could only keep his helium liquid for a few hours and over that time there was no measurable drop in field strength. A similar experiment was carried out with better technology in the 1950s and ran for eighteen months with no discernible drop in field and hence in the current.

'High temperature' superconductors

The behaviour of superconductors is remarkable, and potentially extremely useful for enhanced power transmission and producing ultra-powerful magnets. But while our low-temperature technology is far better than it was in Kamleringh Onnes' time, getting down to 4 K or lower is still an arduous task today – we are talking, after all, –269.15°C (–452.47°F). However, experimenters gradually managed to push up the temperature at which superconductivity could happen to temperatures between 20 and 30 K by developing specialist materials to act as the conductor. Still very low temperatures, but ever so slightly more achievable. However, when in the 1950s the basic theory of how superconductivity worked was developed, it seemed likely that this was the end of the road. And such temperatures remained the limit for around 30 years. But the mantra of the good physicist is 'Never say never'.

BCS theory

The explanation of conventional superconductivity would require the development of quantum theory, which was in its infancy when Kamerlingh Onnes made his discovery. Electrical currents in conductors are carried by electrons, which are only very loosely associated with the outside of the atom of a conductor and which can easily be separated to drift through the conductor's lattice of atoms. As those atoms are constantly in motion, even in a solid, it is hard for electrons to pass through without some interaction with the atomic lattice, causing electrical resistance.

Three American physicists – John Bardeen, Leon Cooper and Robert Schrieffer – developed what would become known as BCS theory to explain the phenomenon of superconductivity. Cooper had already described potential low-temperature interactions between pairs of electrons (imaginatively known as Cooper pairs), which can act as if they were a single particle, linked together by vibrations in the crystal lattice of the conductor. However, these vibrations would also tend to quickly break up the pairs. But at extremely low temperatures, electron pairs, which being quantum entities don't have exact locations, can overlap sufficiently to become a single entity known as a condensate. This gives them the ability to ignore the lattice vibrations and float though the lattice as if it were not there, producing zero resistance.

In March 1987, a team announced the first example of superconductivity occurring at 90 K. This was achieved with the unlikely-seeming candidate of a ceramic. A ceramic is a non-metallic crystalline material, and the most familiar ceramics are good insulators. Look at the insulators used to separate the 25,000-volt cables of an electric railway line, or the six-figure voltages of overhead power lines from their support towers and you will usually see objects with multiple ridges that are ceramic insulators.

These conventional ceramics (and the ones used in pottery) are typically silicates, but the high-temperature superconductor was a more complex structure featuring barium, copper, yttrium and oxygen. It wasn't clear to anyone how the new superconductor worked – the theory described above could not be responsible for this type of superconductivity. As a result, rather than trying to construct new superconducting materials based on theory, experimenters had to rely on trying all kinds of mixes in the attempt to find a substance that would act as a superconductor at even higher temperatures. Within a year, superconductivity was being reported at 125 K by substituting strontium and bismuth for some of the original elements.

Although there have since been a number of reports of superconductivity being observed at much higher temperatures – coming close to room temperature – none has so far proved reproducible. However, efforts continue to this day to produce a room-temperature superconductor, aided by a better understanding of the strange structures in the ceramic superconductors that seem to be responsible for their behaviour, even though the exact mechanism is still not certain.

However, that does not undermine the importance of the new superconducting materials. The liquid helium used in the original experiment remains hard to obtain, expensive to produce and tricky to use. However, liquid nitrogen, which boils at 77 K – so is plenty cold enough for the new superconducting materials – is readily available (it is even used by high-end chefs to instantly chill food), cheap and relatively easy to handle.

Kamerlingh Onnes, the person

Unlike a Newton or an Einstein, Heike Kamerlingh Onnes' life has not been widely documented. However, we can say some things about him as a person. In the early years of the 20th century, there was still considerable deference to social status, yet even by the standards of the time, Kamerlingh Onnes was considered to be old fashioned. He seems to have run his laboratory on near-military lines, and though he had a sizeable staff, his scientific papers were often apparently authored by him alone, as if he were a traditional solo scientist. He was considered paternalistic and overbearing at a time when science was becoming democratised, driven by skill and ability rather than background.

Where at the start of the 19th century, for example, the Manchester chemist John Dalton was unusual in not having a privileged upbringing, by the time of Kamerlingh Onnes' discovery, things had begun to change. Having said that, the Nobel Prize biography of Kamerlingh Onnes says: 'A man of great personal charm and philanthropic humanity, he was very active during and after the First World War in smoothing out political differences between scientists and in succouring starving children in countries suffering from food shortage.' These institutional biographies tend to produce a sanitised view of the individual, but there is not a significant inconsistency between being philanthropic in the wider world and running a lab as a private fiefdom.

It's interesting to contrast the apparent strict hierarchy in the Onnes lab with the situation Danish physicist Niels Bohr found when he joined Ernest Rutherford's laboratory in Manchester in 1912. At Manchester, Bohr said he experienced 'the enthusiasm with which the new prospects for the whole of physics and chemical science, opened by the discovery of the atomic nucleus, were discussed in the spring of 1912 among the pupils of Rutherford'. By contrast with the seemingly cold, detached reception Bohr had discovered when working with English physicist J.J. Thomson, which appears to have been more like that of Kamerlingh Onnes, Bohr found the approach taken in Manchester much more conducive to developing his ideas. Every afternoon there was an opportunity to discuss new ideas over tea and cake, informal get-togethers often presided over by Rutherford, alongside his more formal Friday afternoon colloquia. The big difference for Bohr seems to have been this more collaborative, information-sharing approach. In Manchester, in the heady atmosphere of Rutherford's lab, the quantum atom was conceived. It's hard to imagine that one of Kamerlingh Onnes' assistants would have felt that he had the same freedom.

Life changers

MRI

The MRI scanner, which was one of the most important additions to medical diagnostics in the 20th century, requires extremely strong magnets to flip the magnetic alignment of protons in the body. Such powerful magnetic fields were only made possible with the development of superconducting magnets.

Maglev

Another application of powerful magnetic fields is the magnetic levitation or 'maglev' train. These use magnetic fields, based on superconducting magnets, to float the train above the track, making it possible to achieve speeds beyond anything practical on a traditional railway line. Experimental maglev trains have achieved speeds of over 600 kilometres (370 miles) per hour. At the time of writing there are only a handful of short-range maglev trains fully operational – for example, the 30-kilometre (19-mile) link of Shanghai Airport with the city, which reaches 430 kilometres (270 miles) per hour – but other routes are expected.

More to come

In the examples above, we are still only scratching the surface of the potential of superconductivity. As we have seen, the temperatures at which superconductors operate have increased far further than was ever expected, and experiments into the possibility of a room-temperature superconductor continue. Not only do superconductors enable the construction of extremely powerful magnets, such as those used in MRI, maglev and particle accelerators such as the Large Hadron Collider at CERN, they mean that electrical currents can be carried without loss to heat. If room-temperature superconductors were available, they could transform power distribution in a world that makes increasing use of electricity as we move away from fossil fuels. Equally they could allow more complex electronic circuits to be packed into the same space, as they are often limited by the heat generated by resistance in the circuitry.

Tuesday, 16 December 1947

John Bardeen and Walter Brattain – First demonstration of a working transistor

In the preceding chapters, we have seen the development of concepts in physics that would have significant practical value – this, however, is the first of three days in which it was an *application* of fundamental physics that resulted in something new that has transformed the world. The changing nature of research is also reflected in the move from individuals making the breakthrough to teams, where there is less benefit to understanding the development in exploring individual lives in any detail. A name that is often associated with the development of the transistor is William Shockley, but it was Bardeen and Brattain, working for Shockley in the melting pot of minds that was Bell Labs, whose day came in 1947 with the first working transistor. This was not the beginning of electronic devices, but up to this point electronics was primarily a clumsy technology, limited by the capabilities of thermionic valves (vacuum tubes). With the transistor, electronics could truly begin to feature in all parts of our lives. Bardeen and Brattain's relationship with Shockley was not always an easy one – the birth pangs of the transistor were both human and fascinating.

The year 1947

This year saw the first televised sessions of the US Congress, post-Second World War peace treaties signed in Paris, the demonstration of the first Polaroid camera, the launch of the International Monetary Fund, a cargo of fertiliser exploding in Texas, killing over 500 and destroying 20 blocks of the city, King Frederik IX taking the throne of Denmark, the first Ferrari car launched, Anne Frank's diary published, the Roswell 'UFO' incident, Pakistan and India gaining independence, the marriage of Princess Elizabeth of the UK to Philip Mountbatten and the first commercial microwave going on sale. Among those born this year were English musician David Bowie, Japanese prime minister Yukio Hatoyama, Princess Christina of the Netherlands, English musician Elton John, Indian-born British author Salman Rushdie, English musician Brian May, Camilla Duchess of Cornwall, Austrian-born American actor Arnold Schwarzenegger, English racing driver James Hunt, American writer Stephen King and American politician Hillary Clinton. Deaths included American gangster Al Capone, American store owner Harry Selfridge, German physicist Max Planck, British prime minister Stanley Baldwin and King Victor Emmanuel III of Italy.

John Bardeen in a nutshell

Physicist

Legacy: superconductivity theory, electronics

Born 23 May 1908 in Madison, Wisconsin, USA
Educated: University of Wisconsin and Princeton University
Married Jane Maxwell, 1938
Joined Bell Labs, 1945

John Bardeen

Professor of Electrical Engineering and Physics, University of Illinois at
  Urbana-Champaign, 1951

Nobel Prize in Physics, 1956

Nobel Prize in Physics, 1972

Died 30 January 1991 in Boston, Massachusetts, USA, aged 82

Walter Brattain in a nutshell

Physicist

Legacy: electronics

Walter Brattain

Born 10 February 1902 in Xiamen, Fujian Province,
  China

Educated: Whitman College, Universities of Oregon
  and Minnesota

Joined Bell Labs, 1929

Married Karen Gilmore, 1935

Visiting lecturer at Harvard University, 1952

Nobel Prize in Physics, 1956

Married Emma Jane Miller, 1958

Visiting lecturer, then professor at Whitman College, 1962–76

Died 13 October 1987 in Seattle, Washington, USA, aged 85

Controlling electrons

As a result of Michael Faraday's 'experimental researches' on Day 2,
electricity had been transformed from an interesting topic for study
to a practical and useful means of distributing energy that rapidly
displaced some uses of fossil fuels. This is a process that is still underway, as we see these fuels being phased out in vehicles, heating and

industry because of their impact on climate change. However, electricity as a power source was not the only way that a flow of electrons would prove to have practical value.

Back in the 1830s, Faraday had noticed that putting an electrical voltage across a pair of plates in a glass tube with reduced air pressure produced a strange glow, which as we have seen was investigated in more depth by English physicist William Crookes, who devised the cathode ray tube. In Day 5, we saw that experiments with these devices resulted in the discovery of X-rays, produced when what was discovered to be a flow of electrons through the evacuated tube hit a metal plate at high velocity. (It was through the use of a Crookes tube that J.J. Thomson discovered the electron.)

As various scientists and inventors, notably Thomas Edison, experimented with Crookes tubes, they discovered that using different charged plates within the tube could affect the flow of electrons. In 1904, English physicist John Fleming discovered a practical use for these effects. He produced a device that consisted of a wire which was heated by an electrical current and a metal plate (later supplanted by a metal cylinder around the wire). The heated wire freed up electrons, ready to conduct a current.

The result of this setup was that a current would flow in one direction from the wire to the plate when the plate was positively charged, but would not flow in the other direction, because the plate was not heated, so did not produce free electrons. At the time, early radio receivers used a semiconductor device known as a 'cat's whisker' to extract the signal from a radio wave, but these were difficult to use and constantly needed adjustment to keep them working. Fleming's device, which became known as a Fleming valve or oscillation valve, had the same effect as the cat's whisker, but was much more stable.

What Fleming had made was what is now called a diode (because it has two electrodes) – an electronic component that allows current to flow in one direction but not the other. The general type of device of which this was the first example was known as a thermionic valve in the UK – as it used a heated wire to produce electrons to make it act like a one-way valve – and a vacuum tube in the US, after the evacuated glass tube that surrounded the electrodes.

Such electronic valves were taken to the next level in 1907 by American inventor Lee de Forest. He placed a third electrode, in the form of a wire grid, between the heated cathode and the anode. When he applied an electrical voltage to this grid it changed the amount of current that would flow through the valve. Small changes in the voltage on the grid resulted in very large changes in voltage between the anode and cathode. De Forest's 'Audion' would become known as a triode, and would soon be the backbone of electronic developments.

Central to the value of the triode, was the way that a small electrical current could control the flow of a larger one. This could be used in two ways. The most important for the general public at the time was that the triode acted as an amplifier. A small signal, such as that picked up from a radio receiver or from the needle of a gramophone, could be fed to the grid and made into a strong enough signal to power a loudspeaker.

For the builders of early computers in the 1940s, the triode had another function – it could act as a switch, with the small current turning the large one on and off. Switches are at the heart of the logic circuits required to build a computer, and the computers of the day were limited by the number of valves that could be crammed into one place. Not only were they large, valves ran hot – early electronic computers pumped out a huge amount of heat. The valves were also fragile, and had relatively short lives.

The scale and problems generated by the use of valves as switches in computers can be seen in the construction of ENIAC, the first fully programmable electronic computer. It was beaten by nearly two years as an electronic computer by the Colossus machines at Britain's wartime Bletchley Park codebreaking centre, but when ENIAC went live at the end of 1945 it was arguably the first true ancestor of modern computers in being truly general purpose. However, being dependent on valves made ENIAC anything but user friendly.

This monster machine featured a total of over 17,000 valves, filled a room 30 metres (100 feet) long, weighed 27 tonnes and required 150 kilowatts of electricity to run it. Most of that electrical energy went to producing heat – bear in mind the cathode of each valve is essentially a small electric heater – meaning that the room housing ENIAC needed constant cooling. Like overloaded incandescent light bulbs, valves also regularly burned out. ENIAC never managed to run more than five days without breaking down and a typical time between failures was two days. When valves failed, engineers were then faced with an electronic component version of 'Where's Wally?', attempting to find a burned-out tube among its 17,000 companions.

It is notable that when American science fiction author James Blish described exploration of the atmosphere of Jupiter using a construct called the Bridge in his novel *They Shall Have Stars*, he remarked, 'there was no electronic device anywhere on the Bridge since it was impossible to maintain a vacuum on Jupiter.' The assumption Blish made was that the extreme pressure of the atmosphere would crush any valves. *They Shall Have Stars* was published in 1956, when the answer to this and the other problems of vacuum tubes was already known, if not yet widely in use.

The power of semiconduction

Despite their fragility, valves initially proved a huge success. Vast numbers of households were equipped with a 'wireless' – a radio set, which used valves for demodulating* the incoming signal and for amplification, reflected in the need to wait for these old radios to 'warm up' before using them. But few households possessed more than one or two electronic devices. The invention that would take electronics from clumsy fragility to reliable miniaturisation and into use in a plethora of applications was the result of work at Bell Labs in America. This was the research arm of the telecommunications giant AT&T, which was originally, as American Telephone and Telegraph Company, a subsidiary of the Bell Telephone Company, established by telephone pioneer Alexander Graham Bell (or, strictly, by Bell's father-in-law) in 1877.

John Bardeen and Walter Brattain had been working at Bell Labs on a device that could replace that most important of thermionic valves, the triode. What the team hoped was to replace the function of the triode with a small piece of semiconductor, a fraction of the size of a valve and giving off far less heat. Bardeen was the theoretician who understood the quantum physics at the heart of this new device, while Brattain was the engineer responsible for making it happen.

The importance of understanding quantum physics here cannot be overstressed. When Lee de Forest made a triode, he admitted that he had no idea how it worked. He was an old-school inventor, pure and simple. Using valves (as the name suggests) was like plumbing

---

* Radio signals are transmitted by 'modulating' a carrier wave, which involves imposing the signal on either the size of the wave (amplitude modulation, or AM), or the frequency of the wave (frequency modulation, or FM). Early radio was AM: the signal was removed by 'demodulating' the wave, stripping the variations by only taking the variations in signal from one side of the carrier.

with electrons. However, making use of semiconductor devices was only possible with an understanding of the strange behaviour of quantum particles such as electrons, and the quantum structure of materials. The electronics game had moved from being the province of 'suck it and see' inventors to the theory-driven world of the physicist.

The quantum

We have already seen a number of cases in earlier days when quantum physics played its part in the new physics, but the transistor was the first example where having a sophisticated understanding of the quantum nature of the very small made the breakthrough possible.

The 'quantum' in quantum physics refers to an amount of something, but the key to understanding its importance is that at the level of very small particles, such as electrons, atoms and photons of light, what appear to be continuous phenomena are in reality broken up into small chunks. So, for example, light, which had been thought to be a wave, can be described as a flow of individual photon particles.

Of itself this isn't revolutionary – but the implications of this change of understanding was that quantum particles behave in a totally different way to the familiar objects we can see in the world around us. Quantum particles may often be portrayed as if they were tiny balls, but in reality, they exist as fuzzy clouds of probability where their very location can only be pinned down at the point in time that they interact with something else. It was this understanding of the strange behaviour of quantum particles that made the development of the transistor possible.

American physicist William Shockley shared the Nobel Prize for this development with Bardeen and Brattain, and arguably could have been listed along with the other two men as a key protagonist of this day. Shockley did set the project's goal of producing a semiconductor equivalent of a triode – but it was Bardeen and Brattain (aided by a number of other Bell Labs staff) who made it happen. Shockley himself emphasised this, writing later, 'My elation at the group's success was tempered by not being one of the inventors.'

This was no overnight development. Semiconductors had been in use for decades in devices such as radio receivers where the so-called cat's whisker receiver made use of a semiconductor crystal – often lead sulfide – that acted as a diode. As noted previously, this is a component that only allows current to flow in one direction, and as we have seen, the result was to demodulate the incoming signal. However, such semiconductor devices could be fiddly to use, which is why they were initially often replaced with diodes based on valves.

Semiconductors

Semiconductors are elements that sit between metals – such as iron or copper – and non-metals – such as oxygen or sulfur – on the periodic table. On the whole, metals conduct electricity, while non-metals don't. (The significant exception here is carbon, which is both a non-metal and a good conductor due to its unusual physical properties.) Semiconductors, such as silicon, selenium and germanium, sound as if they should be materials that conduct inefficiently, but in reality they are substances that conduct electricity in some circumstances and don't in others, making them interesting materials for anyone who wants to be able to control the flow of electrons, the role of electronics.

The usual means by which a substance conducts electricity is by a flow of electrons through the material. For this to happen, it has to be possible to free electrons from the grip of atoms. When we look at a single atom, the electrons around the nucleus exist in clouds of probability known as orbitals.

But when atoms are in close proximity in some materials, the outer orbitals effectively run together, eventually forming a near-continuous band that enables electrons to move relatively freely through the material.

In a non-conductor there is a big gap between the normal atomic orbitals and such a 'conduction band', where in a metal there is little or no gap. Semiconductors typically have a narrow gap, which can be bridged by some form of external stimulation, in some cases from incoming light, or from the addition of small amounts of other materials, a process known as doping.

When electrons in a semiconductor are boosted in energy but stay just below the conduction band, in what's known as the valence band, they flow in the opposite direction to the main electrical current, carrying with them any gaps that existed between electrons. These gaps are known as holes and can be treated as if they were particles in their own right. So, in a semiconductor that has electricity flowing through it in the conduction band there will typically be a flow of electrons in one direction and a flow of holes in the opposite direction. The role of doping in semiconductors is to add small amounts of other elements either to provide extra electrons (a so-called n-type agent) or extra holes (a p-type agent). Bardeen and Brattain were working with a semiconductor using the element germanium. Their experimental device was nothing like the tiny component we now think of as a transistor.

Into the solid state: the 1947 day

On a metal base electrode was seated a grey crystal of germanium. This had been treated with doping agents so that the top layer of the germanium had an excess of holes – this top layer was p-type. The rest of the germanium had an excess of electrons – it was n-type.

Above the germanium was a triangular plastic wedge, which had gold foil wrapped around it. Brattain had sliced through the foil at the downward-facing point of the triangle, so the gold on the two sides leading down to the point could separately conduct electricity.

The gold-covered triangle was pressed down onto the germanium by a spring. The result was that the two strips of foil acted as a pair of electrodes with a very narrow gap between them, bridged by the top surface of the germanium. When a small current was passed between one of the gold electrodes through the semiconductor to the base electrode, it controlled the flow of a much larger current between the other gold electrode and the base.

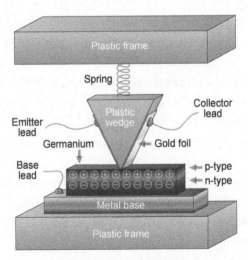

The experimental transistor
assembled by Bardeen and Brattain.

Brattain and Bardeen's success seems to have pushed Shockley over the edge. He had been working on failed attempts to produce a solid-state* triode using a different 'field effect' mechanism (of which more later) for a number of years and he was Bardeen and Brattain's manager. Shortly after the successful demonstration of the transistor, Shockley told Bardeen and Brattain that he felt that he alone should be named on a patent for the concept as he had been working on solid-state valves earlier, even though his had failed. Brattain later said that Shockley had told them, 'sometimes the people who do the work don't get the credit for it.'

The pair were stunned by Shockley's gambit. Bardeen, who was a man of few words, said little, but Brattain shouted, 'Oh hell, Shockley, there's enough glory in this for everybody!' Despite their protests, Shockley started applying for a patent, and looked likely to be successful until earlier patents were discovered from a Austro-Hungarian-born American physicist called Julius Lilienfeld. Although Lilienfeld had not constructed a successful device, his design was fairly similar to Shockley's early ideas, on which Shockley had hoped to base his patent claim. As a result, the Bell Labs lawyers regrouped and based the patent solely on Bardeen and Brattain's work, naming them – an approach that was sufficiently different to avoid any conflict with Lilienfeld's patent. The table was turned and Shockley was excluded.

Shockley's disappointment mollified Bardeen and Brattain, and also seems to have boosted his creative drive. The original so-called point-contact approach would be the basis of the first commercially available transistors, but the design would not last long, as it was just as messy as it sounds. Shockley came up with a totally new design, producing the first junction transistor in June 1948. These were

---

* Solid state refers to the device being dependent on a solid semiconductor rather than a vacuum tube.

doping sandwiches. Instead of having two layers, they had three, with either p-type sandwiched between n-types, or with n-type sandwiched between p-types. Variants on this type of transistor would dominate the field for two decades. This development occurred the same month that the transistor was announced to the world.

The name 'transistor' was the winner of a ballot within Bell Labs (in the days when asking for suggestions did not result in names like Boaty McBoatface). There's no doubt that John Pierce's entry of 'transistor' was catchier than some of the suggestions, such as surface states triode or semiconductor triode. There was another snappy contender in 'iotatron', which played on the fondness of the period to give scientific devices a 'tron' ending (think cyclotron from the 1930s and synchrotron from the 1940s), but transistor had the advantage of an ending reminiscent of existing electronic components: resistors, capacitors, varistors and thermistors.

An early change in the development of the transistor was to move to silicon as a semiconductor, rather than germanium, which was more expensive than silicon and harder to work with. Both semiconductors were familiar to those who had worked on radar during the war. Although germanium had proved easiest to get working in a point-contact transistor, silicon would prove far more effective in the more sophisticated designs. The first silicon transistors were produced in 1954, and rapidly displaced germanium. However, there was one other change that would be needed to get to our current position – a new technology called metal-oxide semiconductors, which made it possible to construct the field-effect transistor, the approach that Shockley had worked on for so long, but which had proved impossible to make without the metal-oxide approach.

With the development of the MOSFET (metal-oxide semiconductor field-effect transistor) in 1959, what had been a fairly clumsy piece of technology a centimetre or two across would take the initial miniaturisation possible in moving from valves to transistors to the next level. The metal-oxide semiconductor part refers to the use of a thin oxide layer on top of a slice of silicon to produce electronic components in a thin layer, which makes it possible to construct the integrated circuit chips that are at the heart of most modern electronics.

The field-effect design reflects a different approach to the control of a current in the semiconductor. Here, the electrical field from a separate electrode called a gate is used to influence the flow in the semiconductor. This proved far more difficult than was first hoped, as quantum effects in a semiconductor tended to exclude the electrical field, but was eventually made possible after an accident in 1955 that left a layer of silicon dioxide on top of a silicon wafer, which prevented this effect from occurring.

Life changers
Electronics
As we have seen, electronics date back to the early years of the 20th century, but it was only with the development of the transistor that it would be possible for electronic devices to be robust, miniaturised and versatile enough to have the central role they have today. The typical 21st-century house will contain hundreds of electronic devices, from the sophistication of mobile phones, through the engine control systems of cars to the simple control mechanisms of toasters and kitchen timers.

Microchips
For the first twenty years or so of their existence, transistors were typically as large as a fingernail. However, the ability to produce

MOSFET transistors and then integrated circuits took the deployment of transistors to a whole new level. Each transistor takes the role of one of the valves in a computer like ENIAC. That had around 17,000 such valves. A typical modern computer processor chip has several *billion* transistors. Bearing in mind that there are billions of mobile phones in the world, each a computer in its own right, and billions of other computers, we are looking at billions of billions of transistors in these chips – without considering all the secondary chips, such as graphic controllers, and the control chips in lesser devices.

Wednesday, 8 August 1962

James R. Biard and Gary Pittman – Patent filed for light emitting diode

The light emitting diode (LED) might seem like the laser's poor cousin, but in terms of impact on our lives, it easily outshines the laser – and in practice, the majority of lasers we use are based on designs that are similar to LEDs. While the laser's most familiar domestic applications such as the CD and DVD are rapidly being made redundant by the development of Day 10 (see below), the LED has gone from strength to strength, giving us long-life, low-energy lighting and the essential illumination for the screens that have come to be so important to us. Here it is far less clear than was the case with the transistor who the central characters in the story were. Yet, with Biard and Pittman's patent, a new approach to one of our oldest technological requirements – artificial light – was made commercial, in a story that had a long gestation.

The year 1962

This year saw Western Samoa, Rwanda, Burundi, Jamaica, Trinidad and Tobago, and Uganda gain independence, John Glenn become the first American to orbit the Earth, the new Coventry Cathedral consecrated in England, Rachel Carson's book *Silent Spring* published, the first Walmart store opened, the launch of the first commercial communication satellite, Telstar, the Beatles' first single and the first James Bond film released, the Cuban missile crisis, the Sino-Indian war, and the agreement to build Concorde signed. Births this year included English author Malorie Blackman, American writer David Foster Wallace, American musician Jon Bon Jovi, English rower Steve Redgrave, Princess Astrid of Belgium, English politician Keir Starmer, Australian film director Baz Luhrmann and American actress Jodie Foster. Deaths included English writer Vita Sackville-West, English composer John Ireland, English statistician Ronald Fisher, American actress Marilyn Monroe, American first lady Eleanor Roosevelt, Danish physicist Niels Bohr and Queen Wilhelmina of the Netherlands.

James R. Biard in a nutshell

Electrical engineer

Legacy: LEDs

Born 20 May 1931 in Paris, Texas, USA
Educated: Texas A&M University
Married Amelia Clark, 1952
Joined Texas Instruments, 1957
Joined Spectronics, 1967
Joined Honeywell, 1978

Gary Pittman in a nutshell

Chemist and electrical engineer

Legacy: LEDs

Born 20 October 1930 in Wellington, Kansas, USA

Educated: Southern Methodist University

Joined Texas Instruments, 1953

Joined Spectronics, 1969

Joined Honeywell, 1978

Died 28 October 2013 in Richardson, Texas, USA, aged 83

Fractured beginnings

Up to this point in our ten days there has been little doubt over which was the key date in that particular story. With light emitting diodes – LEDs – the light sources that have revolutionised artificial lighting, things are far less clear. The phenomenon that is utilised in an LED was first observed in 1907. There were a number of reports of LED-style lighting in the intervening period, including a patent for a green LED from RCA in 1958 which seemed not to be followed through, before arriving at our key date of 8 August 1962, when Biard and Pittman filed a patent.

However, this first LED to be produced with a commercial potential emitted light in the near-infrared – it was not quite visible. The first to produce a light we could see was demonstrated a few months later – this was red in colour, and would be widely used for indicator lights. Red LEDs really started to make their mark when Hewlett-Packard (HP) produced an LED display that would be used in calculators and digital watches. A yellow LED followed in 1972.

By now, LEDs were mainstream as low-energy indicators – yet it would take decades for LEDs to assume their current role as mainstream lighting. This was only made possible with the development of blue LEDs (which won their inventors the Nobel Prize in Physics). Relatively soon after, the true white LED was produced, based on the blue LED, and the incandescent light bulb was doomed.

Any one of these points in history could have been picked to highlight, which emphasises how many modern breakthroughs are not simple, one-off moments. Although I have chosen to emphasise the 1962 date, as the point where the LED went from a technical novelty to a practical product, all these contributions to modern lighting will come into the story. Equally, although Biard and Pittman are major figures in the story, they are far less known as individuals than most of the key figures we have met. It is their technology that will tell their story, rather than any insight into their lives.

Light in the darkness

For the vast majority of human existence, the only source of artificial light was a flame. There is some evidence for controlled use of fire that predates *Homo sapiens* – earlier hominids had certainly made use of naturally occurring fires for some considerable time – but from the flowering of human inventiveness, which seems to have occurred between 70,000 and 100,000 years ago, fire became a widespread tool that could not only make food safer and more palatable but provide a defence and, crucially, light on a dark night.

In a world where more recent forms of artificial light are everywhere, to the extent that we need to have specially protected regions where it still gets truly dark enough at night for astronomers and amateur stargazers to properly see the night sky, it seems remarkable that this early dependence on flame for lighting would continue well into the 20th century. Indoor gas lighting may have been largely phased out by the first decades of the 20th century, but my hometown railway station still had gas lights on the platforms in the late 1960s.

However, electric lighting did become dominant, primarily reliant on incandescent bulbs, lit by heating a wire until it glowed white hot. Other sources existed – most commonly based on electrical discharges in low-pressure tubes containing mercury vapour, which produced ultraviolet light that then stimulated a fluorescent material to give off visible light. These fluorescent lights and incandescent bulbs were relatively cheap and reasonably long lasting, but used a considerable amount of energy for the amount of light given off. This was particularly the case with incandescent lights, which pumped out significantly more energy as heat than was used in producing light.

Growing incandescent

The history of the incandescent light bulb shows both the power of advertising and the danger of putting too much trust on the 'lone genius' model of invention – a particularly important concern when later looking at light emitting diodes, which do not have a single, clear inventor.

Ask a person on the street who invented the light bulb and they will almost certainly point to Thomas Edison. Edison was, without doubt, a great inventor. Neuroscientist Simon Baron Cohen has suggested that Edison had an unusual ability that resulted from being reasonably far up the autism spectrum in being compelled to repeatedly test patterns over and over again. He tried vast numbers of potential filaments for a light bulb, but they repeatedly burned out in a short space of time.

However, Edison's was not the first electric light bulb by any means – and not even the first capable of commercial production. The English inventor Joseph Swan produced a working light bulb, based on a carbon filament (as was Edison's) eight months earlier in 1879 than did Edison. But unlike Swan, Edison was a cut-throat businessman. Infamously, when attempting to have his

DC electrical system favoured over his rival's AC system, he had demonstrated how dangerous the AC system was by using it to electrocute an elephant. Edison attempted to push aside Swan's priority with a patent infringement case, but lost. Edison was forced to set up a joint company, the Edison and Swan United Electric Light Company to produce their invention.

At the most fundamental level, every light source except nuclear processes works the same way. The electrons that exist in a fuzzy layer around the outside of atoms can be stimulated to jump up to a higher energy level than the one they usually occupy. Such a position is unstable: soon after, the electrons tend to drop back down, and give off their excess energy in the form of a photon of light. The only significant difference between our artificial light sources is how the electrons are stimulated in the first place.

In a flame, chemical energy that is released when the fuel burns gives the electrons the desired boost. In traditional electric light bulbs, it is the electrical current impacting on the atoms in the filament or vapour that stimulates the electrons. In principle, the term 'electro-luminescence' (which we are about to explore) could be applied to such electric light sources. However, it is reserved in practice for a very specific way that electrons lose their energy when interacting with the phenomenon given the uninspiring name of 'holes'.

The glow of holes

We met holes in Day 8 as one of the features that semiconductors bring to electronics. Bearing in mind that a hole is essentially a space where an electron can fit but isn't present, it is no surprise that an electron can drop down from the conduction band into a hole. As this is a reduction in energy for the electron, the outcome is the emission of a photon of light. And it is the production of light by electrons dropping into such holes that is the definition of electroluminescence.

The phenomenon was first observed in 1907 by an English electrical engineer, Henry Round, who worked for the Marconi Company, then world leaders in radio. As we saw in Day 8, at the time, radio receivers often used something known as a cat's whisker, which was a semiconductor that acted as a diode. Round was a prolific inventor who would end up with 117 patents under his belt.

While Round was experimenting with cat's whiskers, he noticed that some glowed with light when in use. Round wrote a technical letter to *Electrical World*, noting that 'On applying a potential of 10 volts between two points on a crystal of carborundum [silicon carbide], the crystal gave out a yellowish light. Only one or two specimens could be found which gave a bright glow on such a low voltage, but with 110 volts a large number could be found to glow.' Round seems not to have continued with electroluminescence, but between the 1920s and 1930s, a Russian engineer, Oleg Losev, carried experiments significantly further, though as yet the theory behind the phenomenon was still unclear.

It was only in the late 1950s, with the growing understanding of solid-state electronics that had led to the development of the transistor, that LEDs would be seriously considered. For many years, electroluminescence would be an interesting but mostly useless physical phenomenon. But it would turn out that this strange semiconductor behaviour would accidentally spark the LED revolution.

Lasers and their spinoffs

Before we reach the LED, it's worth establishing what its powerful but less widely used big brother is all about. A laser, short for Light

Amplification by Stimulated Emission of Radiation is a device that achieves an unusual kind of light, produced by effectively double-loading the usual mechanism for producing light. A beam of light is passed through a material, and it is the light energy that is used to push electrons up to a higher level.

It might seem that this is pointless – a photon of light is being used to stimulate an electron, which then produces a photon of light. However, the trick is that rather than wait for the electron to drop of its own accord, a second photon is used to trigger the drop. As a result, the beam of light carrying these second photons is amplified – one photon goes in, two come out.

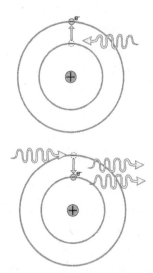

Upper image: *An electron absorbs a photon and jumps up in energy.*
Lower image: *A second photon triggers the release of energy, emitting two photons.*

It was Albert Einstein (again) who speculated that such a stimulated emission of radiation would be possible, back in 1916. By 1954,

Russian physicists Alexander Prokhorov and Nikolay Basov had published a paper on using this mechanism to produce a device that amplified microwaves – electromagnetic radiation that is still made up of photons, but of lower energy than those of visible light. The same year as Prokhorov and Basov published their paper, American physicist Charles Townes independently produced such a device, based on stimulating emission of radiation in a ruby crystal, which he named a maser.

Masers were interesting, and useful in the communications field where Townes worked, but the real prize was to get the same phenomenon working for visible light. A number of labs around the world were working on this, with Art Schawlow, working with his former boss Townes at AT&T's Bell Labs, among the leaders of the pack, along with a former graduate student of Townes at Columbia University called Gordon Gould, who had got a job at defence contractor TRG. By the end of 1958 the race was hotting up. Townes and Schawlow had filed a patent for what they dubbed an optical maser, while Gould (who came up with the term laser) and TRG had asked for a $300,000 grant from the government's Advanced Research Projects Agency, only to be awarded nearly a million dollars.

It was at this point that developments took on the aspect of a farce. Because it was a defence contract, Gould and his co-workers had to undergo security clearance. This was in an America that was deeply suspicious of any hint of communism. Gould had dabbled in left-wing politics as a youth. When this was combined with the fact that he and his wife had lived together before marriage, and the problem that two of Gould's referees for security clearance had beards, which apparently made them potential subversives, his clearance was turned down.

Not only was Gould not allowed in his own lab, he wasn't even allowed to read his own notebooks, as his work was classified and he didn't have clearance. Meanwhile, his competitors at Bell Labs had their own problems. They had toyed with using a solid-state material for the laser, such as the synthetic rubies used in masers. But these had been dismissed as too inefficient at visible light frequencies. Instead, they were concentrating on using gases and metal vapours, which seemed better theoretically, but were trickier to handle and dumped on Schawlow a whole host of practical issues to overcome.

In the meantime, an electrical engineer with a physics PhD working at the Hughes Aircraft Corporation was also undertaking a smaller-scale attempt on the laser. Theodore Maiman had already worked with rubies on a maser project and felt that it should be possible to also use them in a laser. He was suspicious of the calculations that Schawlow had used to show that rubies would be too inefficient to work – and experimentally discovered that rubies were about 70 times better than Schawlow had mistakenly thought.

This didn't mean that Maiman had everything cracked. To provide the stimulation he needed a very bright light source, but the obvious solution, arc lamps, were so hot that they destroyed the ruby crystal. By luck, Maiman's assistant, Charlie Asawa, had a friend who was into photography and who had recently bought a new product in the photographer's armoury, an electronic flash, replacing the old one-shot flashbulbs. Maiman managed to adapt a spiral flash tube to fit around a cylindrical ruby with mirrored ends.

On 16 May 1960, Maiman produced the first laser beam from his device. Lasers rapidly became the big new thing, not only replacing masers in communications, but finding wider uses because their sheer concentration of in-step 'coherent' light made them potentially extremely powerful.

Coherence

The most significant difference between lasers and LEDs is that laser light is coherent. If we think of light as a wave, in a coherent light source, the waves of light have the same wavelength and move together in step, giving them a collective impact that can be more dramatic than light from an ordinary source – just as a group of people marching in step across a bridge can set up a more powerful resonance in the structure. As we are talking about quantum technology, if we think about light as a stream of photons, in a coherent beam, each photon has the same energy and a property of the photons called their phase, which varies with time, is synchronised.

By comparison, LEDs produce a spectrum of colours, though a significantly narrower spectrum than white light, so we see them as having a specific colour. Their waves or phases are not in step. The difference is due to the different way the light is produced. For LEDs this involves an electron combining with a hole, whereas in a laser a photon interacts with an atomic electron to generate a second photon.

After the ruby laser, a host of other technologies came into play, including Townes' gas laser. However, most of the devices were bulky and required considerable external support in the way of power packs and more. The range of potential applications of the laser would be widened were it possible to use an electroluminescent semiconductor to produce the laser light.

The breakthrough: the 1961 day

One of the semiconductor materials that had shown most promise for light production was gallium arsenide. By September 1961, James Biard and Gary Pittman at Texas Instruments had produced good near-infrared light production from a gallium arsenide tunnel diode. Tunnel diodes had only been invented four years earlier, making use of a quantum effect called tunnelling that allows quantum particles to pass through a barrier as if it were not there.

At the time, all the focus on the use of LEDs – as was the case with the origin of lasers – was for optical signal communication. When Biard and Pittman filed for their patent on 8 August 1962, they were accepted as producing the world's first practically usable light emitting diode – and it was as a signalling device that they imagined it would be used, for which near-infra red was fine.

The same year, a semiconductor laser, or laser diode, was developed. These were first produced at both General Electric and IBM. Like LEDs, laser diodes make use of electroluminescence (and initially contained the same semiconductor materials as early LEDs), but the laser versions have a more complex structure that allows the build-up of stimulated emission. These semiconductor lasers are the ones now found in most of our commercial laser-based technology, from DVDs and CD players to laser pointers and laser printers.

A year after Biard and Pittman's first LED was produced, Nick Holonyak Junior, working for General Electric, demonstrated a red visible light LED. He had hoped to develop a semiconductor laser – and did soon after – but his first attempt was only coherent at ultralow temperatures: at room temperature, it acted as an LED. These LEDs, and the variants that came out over the next ten years or so, proved not to be powerful enough for communications. LEDs lack the coherent nature and potential power of lasers, though they were cheaper and a lot more compact. Instead, they started to be used as indicator lights. By 1969, HP was producing LED displays to show a handful of numbers, based on a variant using gallium arsenide phosphide as its semiconductor.

By comparison with existing light sources, LEDs were very efficient, using a tiny amount of electricity for the light produced. This made them ideal for battery devices such as calculators. However, to make them a viable alternative to conventional lighting, they would have to be increased in power and made capable of producing white light. In theory, white light can be produced from a mix of the three primary colours: red, green and blue. Red and green LEDs existed, but there were no blue options.

This colour limitation was also a limitation for semiconductor lasers. The amount of information that can be stored on an optical disc like a DVD is limited by the wavelength of the light used. The red lasers used on DVDs (and the infrared lasers on CDs) could not pack in as much information as would be possible with a blue laser. It was the development of blue LEDs and laser diodes that made the higher capacity of Blu-ray possible, in the early 21st century a major breakthrough, although the arrival of streaming would mean that the technology became near-obsolete soon after its development.

Blue light blues

The blue LED saw first light in 1972 at Stanford, but at the time it was both weak – and therefore unsuitable to be the basis of a blue semiconductor laser or a light source – and not a fully functional device. The search was on to find a way to produce a commercially packageable high-intensity blue LED. The early devices had used gallium arsenide, and this would be replaced by gallium nitride in the 1980s and 1990s, using a new process that enabled a much better way to grow the crystals.

Probably the biggest issue to be faced was the substrate – the supporting medium on which the gallium nitride crystals rested. In 1986, Isamu Akasaki and Hiroshi Amano in Japan started with a substrate of sapphire, coated with aluminium nitride, on which the gallium nitride was grown. Although we are used to sapphire as an expensive gemstone, in its industrial form it is a cheap aluminium oxide crystal.

In parallel, Shuji Nakamura, also working in Japan, developed an alternative approach where layers of gallium nitride were grown at a mix of low and high temperatures. He was also able to explain the mechanism behind Akasaki and Amano's breakthrough, and to provide a much simpler and cheaper approach to constructing intense blue LEDs. The trio would win the 2014 Nobel Prize in Physics for the development of the blue LED – even though they were later successfully sued for infringement of an earlier patent by Theodore Moustakas of Boston University.

It might not seem clear why the Nobel committee should pick on the blue LED as worthy of a recognition that earlier developments were not. Nobel Prize awards for technology can often seem random. So, for example, the prize was also awarded for the laser – but not to the actual first developer, Maiman, nor to Gould or Schawlow. Instead, it was Townes, Basov and Prokhorov who shared an award that recognised the far less significant maser.

Most of the media coverage of the prize suggested that the importance of the blue LED was that it completed the ability to produce white light now that red, green and blue LEDs were available. In practice, this is rarely a sensible way to produce lighting as the LEDs are hard to combine effectively in a compact source. The approach is used in some colour-changing bulbs, but these tend to be a lot more expensive and less efficient than a single-colour bulb.

Instead, blue LEDs were used to produce a kind of white light by using a yellow phosphor coating, rather like that on a fluorescent light. The intense blue light of the LED is partly transmitted through the coating and partly turned into reds and greens by the phosphor, resulting in a blue-tinted white. However, the result was lighting that was too cold in tint for pleasant household lighting, and initially these bulbs were relatively inefficient. It was only with the introduction of warm white LEDs that modern lighting has been able to shift away from incandescent and the compact fluorescents that had briefly begun to replace them. These newer LEDs have gadolinium added to the phosphor, enabling it to reproduce the warmth of lighting that we expect from sunlight.

It was, then, the blue LED that made it possible for LEDs to take over as a standard means of lighting.

Lighting the world

It's hard to overemphasise the importance of the LED bulb. It has come at a perfect time when the link of energy use to climate change has become crucially important for the world. A traditional incandescent light bulb only turned about 4 per cent of the electrical energy it consumed into light, radiating the rest as heat. By contrast, a modern white LED bulb converts more than 50 per cent of the electrical energy supplied to light. That remarkable improvement in efficiency is not trivial when you consider that between 20 and 30 per cent of worldwide energy consumption used to go to lighting. In 2014, the US Department of Energy estimated that a switch to LED lighting would save an annual 261 terawatt hours, potentially rising to 395 terawatt hours by 2030. This is greater than the entire electricity consumption of the UK.

We are seeing LED lighting taking over far beyond the role of room lighting. Street lamps, which were once based on an electrical discharge through sodium or mercury vapour, are being replaced with lower-energy, higher-intensity LED lighting. Light emitting diodes turn up in traffic lights and have replaced both incandescent lights and some of the halogen bulbs in car lighting. And the ubiquitous LED has also helped in the development of lower-energy, more effective TV screens, computer screens and mobile phone screens.

For many years, screens relied on cathode ray tube technology. A development of the early Crookes tubes that also inspired X-ray devices, cathode ray tubes were both heavy and very bulky, requiring a protrusion behind the screen nearly as big as the screen's width. By contrast, liquid crystal displays (LCDs) can be very thin. But unlike a cathode ray tube, a liquid crystal screen does not give off any light in its own right. It simply controls how much light gets through it. As a result, a screen based on LCDs needs backlighting.

Initially, backlighting was provided by fluorescent lights or electroluminescent panels. These combine the kind of electroluminescent effect used in an LED with a phosphor that glows in a particular colour, often blue. Such panels were often used in cheap liquid crystal display backlights (for example for digital watches), while fluorescent panels were used in LCD computer and TV screens. Now, however, LED backlighting has become the new norm.

Not only does the LED provide a huge improvement in energy efficiency, it also lasts longer than the technology it replaces. Without the damaging heating and cooling of an incandescent light, which gradually burns off part of its filament as vapour until the filament gives way, the LED light has stability and relatively little production of heat. An LED can last up to 100,0000 hours, where a typical incandescent bulb may only stay in one piece for about 1,000 hours. Although compact fluorescents do last longer than incandescent bulbs, they still undergo stresses from high-voltage discharges, so don't last as long as an equivalent LED, with a lifetime of around 10,000 hours. And fluorescents are less efficient too.

Compact fluorescents are also limited in how small they can be made. For decades, fluorescent lights were mostly available in the form of long tubes. It was something of a miracle that they were ever made compact enough to replace a traditional light bulb, usually by turning the tube into a narrow spiral. However, they could never replace spotlight bulbs, which LEDs do with ease. There was also the much-criticised need for compact fluorescents to warm up, taking minutes to reach full brightness, where LEDs come on at full brightness immediately.

We now also see specialist variants of LEDs in organic LEDs or OLEDs. These make use of an organic material (usually polymers or small molecule carbon compounds) to act as the electroluminescent part of the diode. Although not as powerful as conventional LEDs, OLEDs can be made as ultra-thin layers which have a lower voltage requirement, a wide viewing angle and particularly good contrast when compared with a conventional LCD screen with an LED backlight. And as the OLEDs produce light themselves, they don't need another layer. (There are also screens based on conventional LEDs, but they are only really suited to large-scale displays, such as those used at sports stadiums.)

Life changers
LED lighting
For at least 100 years, electric lighting was primarily provided by incandescent bulbs. The introduction of LED lighting has transformed this, both in energy consumption – an essential consideration for climate change as much as economics – and the lifetime of bulbs.

LED screens

As we have seen, LEDs feature as backlights to many LCD screens and in OLED form act as the actual image-producing screen. As with lighting, low power consumption and long life make LEDs ideal.

Solid-state lasers

Although not strictly the same thing, the development of semi-conductor solid-state lasers has been strongly tied to the development of LEDs – without the LED development, we would not have the current generation of these ubiquitous devices in printers, scanners, optical disc devices and the laser rangefinders used by smartphones and self-driving cars.

Wednesday, 1 October 1969

Steve Crocker and Vint Cerf – First link of the internet initiated

It would be hard to overplay the importance that the internet has had in shaping the 21st-century world. It's not just a matter of the World Wide Web, important though this has been, but also the replacement of most of our communication media, from mail to the telephone, and the transformation of TV from a medium scheduled at the whim of the broadcaster to a far more sophisticated facility that allows viewers to launch programmes and films at their convenience. The October 1969 date signalled the arrival of the second node on a communications network that would later span the world. Just as there is no point in having one telephone, a single internet location meant nothing – but with the installation of a second machine, a revolution was underway. As with Days 8 and 9, it is harder to pick out key individuals, but there is no doubting the importance to the internet of two technophiles who had been best friends at school.

The year 1969

This year saw Russian spacecraft Soyuz 4 and 5 dock in space, Richard Nixon become US president, the last public performance of the Beatles, the first flight of a Boeing 747, the introduction of the Harrier Jump Jet, Robin Knox-Johnston perform the first single-handed, non-stop circumnavigation of the world by sail, Charles de Gaulle step down as French president and Georges Pompidou elected, Prince Charles become Prince of Wales, Apollo 11 put the first humans on the Moon, the Woodstock festival, *Monty Python's Flying Circus* air for the first time, the CCD digital camera invented, the first episode of *Sesame Street* broadcast, colour TV begin in the UK and the UNIX computer operating system launched. Among those born this year, German racing driver Michael Schumacher, Welsh actor Michael Sheen, American actress Jennifer Aniston, English fashion designer Alexander McQueen, American musician Mariah Carey and Australian actress Cate Blanchett. Deaths include English actor Boris Karloff, English author John Wyndham, American general and president Dwight D. Eisenhower, American actress Judy Garland, Vietnamese political leader Ho Chi Minh and American author Jack Kerouac.

Steve Crocker in a nutshell

Computer scientist

Legacy: the internet

Born 15 October 1944 in Pasadena,
  California, USA

Educated: University of California, Los Angeles

Research management at DARPA, USC/ISI and
  The Aerospace Corporation, 1970s and 80s

Co-founder of Cybercash, 1994

Range of other roles

Steve Crocker

Vint Cerf in a nutshell

Computer scientist

Legacy: the internet

Born 23 June 1943 in New Haven,
   Connecticut, USA

Educated: Stanford University, University of
   California, Los Angeles

Assistant professor at Stanford, 1972–76

Research management at DARPA, 1973–82

Vice President, MCI, 1994–2005

Range of other roles

Vice President and Chief Internet Evangelist, Google, 2005–

Vint Cerf

Beginnings

What we now think of as a public utility began as a closed system, developed for ARPA, the Advanced Research Projects Agency set up by the US government in 1958 as a reaction to the shockwaves caused in the western world by the USSR's launch of the first satellite, Sputnik 1, the previous year. ARPA (renamed DARPA in 1972, making clear its military connections by adding 'Defense') had the role of kickstarting and funding blue skies research that might not have a direct and immediate military benefit, but that could be used in a defence role in the future.

ARPA had far greater flexibility than might be envisaged for a military organisation, perhaps in part because it was set up by Neil McElroy, a man who had previously dreamed up the soap opera as a way of promoting products on radio and TV. Among the projects ARPA and its successor would help fund were the early version of the

GPS satellite navigation, robotics, lasers, artificial intelligence, micro-chips and powered exoskeletons. But its greatest legacy is the internet. Computing in the mid-1960s, when ARPA became involved, was a totally different world to our current combination of small local computers in anything from desktop machines to phones with large remote facilities linked by a ubiquitous network. Computers were generally unconnected large machines, often the size of a room, requiring a specialist environment.

Most of the input to computers in the 1960s was handled using punched cards – thin sheets of cardboard, roughly the size of a bank-note, marked out with a series of positions where a rectangular hole could be punched. The most common cards had 80 columns and twelve rows. A program and data would be input into a machine as a deck of cards, automatically read. The design was inspired by the similar cards that had been used in Victorian automated Jacquard looms to set up complex weaving patterns. The output from the computers would be produced on long, fan-folded sheets of paper.

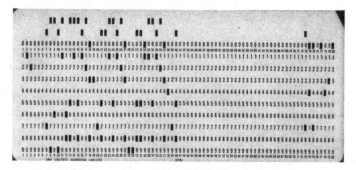

Example of a punched card used to input data into a computer.

Punched cards persisted in computing through to the 1970s. My first experience of computing was at school in Manchester, around

1971. At the time, no British school had its own computer. (My school, the Manchester Grammar School, was the first in the UK to get a computer, in 1977.) We used to set up our cards, knocking out each hole separately using a manual punch, and post the cards to London, where they were run on an Imperial College machine. It took well over a week to get a response (usually that an error had occurred).

However, by the mid-1960s, the most advanced computer facilities also had teletype input where the computer user would type into a device like a typewriter. The program they typed would appear on a piece of paper in front of them, but would also be sent to the computer, which could type back on the same piece of paper, allowing a more interactive style of operation. Such computers, using an operating approach known as 'time sharing' to distinguish them from the one-at-a-time punched card approach (known as 'batch input') allowed several programs to run simultaneously, swapping attention between them.

Whether a computer took its input from a teletype or a card reader, the norm was that the input device would be directly physically connected to the computer. This meant a limit on the ability to access the computer remotely, and also could require an organisation that owned several computers to have a room where a whole bank of different teletypes and card readers were available, each specific to the computers in use. While it was possible to have remote access – for example, there were lines connecting the Pentagon in Washington (where ARPA was based) with computers as far afield as California – these would usually involve dedicated wiring.

The idea for something different was inspired by the psychologist J.C.R. Licklider, who became head of ARPA's Information Processing Techniques Office (IPTO) in 1962. Licklider had a vision of home computers connected together and the development of more intuitive ways of interacting with computers, which he referred to as 'man-computer symbiosis'. One of Licklider's first actions on starting his new job was to reach out to around a dozen computer scientists around the US, who he referred to as the 'Intergalactic Computer Network'. Soon, frustrated by the difficulties of having a whole range of computers that didn't talk to each other, he would send to the group a memo emphasising the benefits of setting up an 'integrated network operation' which would enable the computers to work together, and make them usable without needing to learn a whole new approach for each different computer.

Licklider only stayed in the job a couple of years. In 1965, his successor Bob Taylor suggested an approach to get started on Licklider's vision. Most of the people the IPTO worked with were in universities across the US. They all needed access to computers, which were expensive and in short supply. If it were possible to connect the sites together, so that anyone who was part of the network could connect to any of their computers, the resources could be far more efficiently used. What's more, there was duplication occurring, with a lack of ability to communicate easily between universities. On the back of a single conversation, describing a problem with no specific solution in mind, Taylor was given a million dollars to make it happen.

A different kind of network

Two elements would play a strong role in deciding how that million dollars would be spent. One was the concept of a distributed network. Mathematically speaking, networks are points (known as vertices or nodes) connected together by lines (known as edges). They are a powerful mechanism for analysing and understanding a system. Perhaps the first ever example of network use was the Seven Bridges of Königsberg problem, solved by Swiss mathematician Leonhard Euler in 1736.

The city of Königsberg had seven bridges connecting different parts of the city across the river Pregel. As the river passed through the city it cut off significant pieces of land as islands. The problem involved finding a route taking in each of the areas of land that involved crossing each bridge once only. By abstracting the possible routes as a network with the bridges as edges, Euler was able to show each node (piece of land) was entered by an odd number of edges (bridges). This meant the problem couldn't be solved. As each bit of land was entered and left by a bridge, for such a walk to succeed there had to be an even number of bridges connected to each node that wasn't the start or end of the walk.

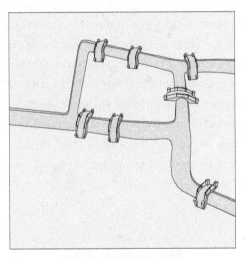

The seven bridges of Königsberg.

Two types of network were already familiar in 1965 from phone and telegraph networks, both forms of hub-and-spoke network. Centralised networks had a single central node, acting as a hub connected to every node. Decentralised networks had multiple central nodes, connected to each other, with each central node acting as a hub for local nodes. But there was a third possibility, which would be the basis for the internet – a distributed network. Here each node is connected to a number of nearby nodes, forming an irregular lattice. Many road systems are decentralised networks. Where a hub-and-spoke network usually provides a single preferred route from A to B, in a decentralised network there are many such possible routes.

Existing phone networks were hub-and-spoke, but the idea of using a decentralised network for communications was suggested in 1960 by Polish-American engineer Paul Baran, working for RAND Corporation, a US think tank widely used by the American military. Baran recommended such a structure to give communications redundancy in case part of the network was damaged during a nuclear war. This certainly was part of the early thinking on the future of communication networks, but rapidly became less significant.

Despite Baran's concept, there was at the time limited interest in distributed networks. Apart from the potential technical difficulties of how best to route a message through such a network, of which more in a moment, distributed networks had not historically been given much consideration because telephone networks were analogue.

Analogue versus digital

The technical distinction between analogue and digital is often that analogue mechanisms tend to be continuous where digital mechanisms are discrete, broken up into chunks – in effect, a digital signal is a quantum version of a traditional analogue signal. Traditional telephone networks sent an electrical wave down the telephone line from A to B. Similarly to the radio transmissions described on page 129, the signal – the voice in a conventional phone call – was carried by modulating the wave. This involved changing the shape of the wave with a second wave that represented the audio signal.

By contrast, a digital signal is simply a string of zeros and ones, where each number is represented by a different electrical voltage. This makes a far simpler square wave pattern, which does not require modulating and demodulating as does an analogue wave. (For this reason, the tendency to refer to the internet routers used to connect us to the internet today as 'modems', which is short for modulator/demodulator, is totally inaccurate.)

Because of their more complex structure, analogue signals deteriorate quickly with the number of connections made in a network, so that passing through many nodes would leave the message unusable. But with a digital network, which only had to distinguish zeros and ones, a distributed network was far more feasible.

Even if data became digital, there was still the issue of how to route messages through the network. Baran came up with the idea of what he called 'message blocks' – small chunks of data that could potentially take a range of routes through the network before being reassembled as a final message. This would enable the network to better manage traffic, rather than having a single line being hogged by a message for the length of time the entire message took to be sent out, as was the case with conventional telephone networks.

We have seen how AT&T's Bell Labs were capable of huge innovation in the invention of the transistor – but AT&T was also an old-fashioned telecoms giant which was used to exerting control over its customers. As was the case with many early telephone companies, it was used to specifying exactly what could be connected to its lines and how the network was used. Like the GPO and its successor BT in the UK, only phones bought from the telecom company could be used on the network. AT&T was simply not prepared to consider a less controlled, distributed approach to its networks, which provided pretty much the only long-distance lines available in the US at the time. After Baran had been working on the concept for five years, it was effectively mothballed in 1965.

The same year, more detail was independently fleshed out on the ability to control the flow of messages through a distributed network. This innovation came not from the might of the US military machine, but from a Welsh physicist. In 1965, Donald Davies, employed by the UK's National Physical Laboratory, started thinking about the mechanics of communicating through a distributed network. After working on his idea part-time for a year, he gave a public lecture proposing a mechanism for data communications which he described as 'packet switching'.

The approach Davies suggested relied on the same kind of blocks of data as Baran's, forwarded from node to node by devices that would be called switches. However, rather than doing this to help the network survive a nuclear attack, Davies believed that such a network would transform computer-based communications to enable remote computer networking and communication between computers – exactly the requirement that ARPA was searching for a solution to. Unlike AT&T, those involved in the British telecoms industry were impressed by this idea and it was Davies' concept and more detailed approach that would feed into the ARPA project.

Initially, when the proposal was put to a 1967 meeting of ARPA principal investigators, the response was mixed. As computer scientist Doug Engelbart (inventor of the computer mouse), who attended, commented: 'One of the first reactions was, "Oh hell, here I've got this time-sharing computer and my resources are scarce as is."' People were worried that opening up their computers to remote users would reduce their own ability to use them. However, this meeting did manage to duplicate one of Davies' ideas, which was to use small intermediate computers on the network to act as switches, taking packets and passing them on to the next node.

ARPANET, as the internet was first called, was by no means the first mechanism for sending information down a line. However, what made it special was the intention, built in from the start, to be flexible. Usually there was a fixed role for a connection, using a formal protocol that was tailored for one specific use. When the US universities involved in the programme were consulted on their requirements, they highlighted two needs that were sufficiently different that the existing approaches could not satisfy them.

The universities wanted the ability to remotely log in to a mainframe computer at another site. But they also wanted to be able to exchange files, so that, for example, a package of data could be sent to another university. These were very different requirements. As Steve Crocker described it, at a meeting in August 1968 of the early university participants, the feeling was that if they were going to all this trouble, they should look for a 'more general framework' to enable the network to be used for a wider range of applications.

Arguably, the most revolutionary aspect of the ARPANET is one that is still shocking today, if frequently unnoticed. When using traditional networks – whether we are talking telephone networks, TV networks or even the early computer networks that the internet would eventually push out of the way, such as CompuServe and AOL – the user paid for the privilege. But because ARPANET was government funded, no one gave any thought to building in a charging mechanism. The ability to bill users simply wasn't part of the system architecture.

Can we login? – the 1969 day

Finally, a connection was wired up between UCLA and Stanford Research Institute (SRI). The role of building the necessary intermediate small computers to control the flow of information packets around the network, then known as IMPs (Interface Message Processors) had been given to a relatively small Cambridge, Massachusetts-based consultancy called Bolt, Beranek and Newman, which achieved remarkable things in a short timescale, given the novelty of the whole concept. The hardware was on its way. But the make or break for ARPANET would be software.

The two individuals whose names are attached to this day, Steve Crocker and Vint Cerf, got to know each other as high school friends with an enthusiasm for science. Together, while still school students, they had got their hands on a computer at UCLA by breaking into the closed computer centre at the weekend and making use of computer down time. After graduating from Stanford with a maths degree, Cerf got a job with IBM in Los Angeles, working on a time-sharing system, but re-joined Crocker at UCLA as postgrad researchers in computer science – just before Crocker moved to MIT. However, the pair were reunited in summer 1968 when Crocker returned to UCLA.

As with LEDs, a large number of individuals could be identified as taking the key step, but Crocker and Cerf were central in the development of the software protocols for the network, working with a number of other graduate students at the participating universities. A network protocol is a kind of universal language, a standard way of gaining access to locations and giving instructions. The underlying protocol of the internet would become known as TCP/IP (Transmission Control Protocol/Internet Protocol). Its role is to enable the splitting of data into packets, addressing them, sending them through the network to their destination and reassembling them as a message.

Protocols, domains and DNS

We don't usually see TCP/IP protocols directly, but another communication protocol, which sits on top of TCP/IP and which is more familiar, is Hypertext Transfer Protocol (HTTP), which allows browsers to make requests of a server, and enables the server to provide the requested information, which is then routed via TCP/IP. (Other internet services, such as email, messaging and video streaming make more direct use of TCP/IP, though they can still have HTTP user interfaces.) HTTP both specifies exactly what a browser wants from a website – for example, a page – and the layout of information on the screen, which is specified via a widely used code known as Hypertext Markup Language or HTML. There are other protocols sitting on top of TCP/IP, for example File Transfer Protocol (FTP) for moving files from computer to computer, and Internet Message Access Protocol (IMAP), used to retrieve email messages from a server.

The domain names used when we access a web page are lower level and provide a relatively friendly way of making a connection to the correct server. Servers are identified by an address known as an IP address, which consists of four numbers, each of which can be between 0 and 255, represented by 8 bits, making 32 bits in all. This means there are total of around 4.3 billion possible addresses. Given the number of devices on the internet, this protocol is undergoing a long-term process of being migrated to a system with 64-bit addresses, allowing for 18 million trillion devices. Computers known as domain name servers translate a familiar URL such as www.brianclegg.net into the appropriate IP address.

It was Crocker who set up the central mechanism of RFCs, or Requests for Comments, used by those in charge of internet protocols as a way of discussing ideas and changes, and setting standards. There have now been thousands of these since 7 April 1969, when RFC 1 was issued, which discussed the basic way two computers would establish a 'handshake' indicating they were in communication.

At the start of September 1969, the first IMP, a modified Honeywell 516 computer, was air-freighted from Boston to Los Angeles and installed down at UCLA. This was a grey, fridge-sized box, weighing about 400kg (900 pounds). It started up without error, but with only one node there was, as yet, no network. The team at UCLA were able to test the link to their computer, a Sigma-7 made by SDS (which later became Xerox), but as yet there was nothing happening.

The second IMP was installed on 1 October at SRI in Stanford, where there was a different and usually incompatible SDS computer, an SDS 940, a computer designed for time-sharing, where the Sigma-7 was originally designed as a batch machine.

The ARPANET was intended to run at 50 kilobits per second. Today it's not unusual for internet connections to the home to run at 1,000 times this rate, while the backbone links are far faster still. As the initial aim was not to communicate on a peer-to-peer basis, but rather to be able to log in to a mainframe computer, the very first characters to go down the line spelled out the command LOGIN. At least, they should have done.

Charley Kline, an undergraduate at UCLA, had the honour of typing those first characters. To check on what was happening, he was also in contact with his opposite junior number at SRI, Bill Duvall, by telephone, checking that each character arrived as he typed it. Kline got as far as LO... and as he typed G, the SRI system crashed. The reason turned out to be some over-smart programming. Because there was no other command at this stage of the interaction that started with the letters LO, the SRI system automatically added the GIN; this was sent back via a program that was only expecting to receive one character at a time and promptly crashed.

This problem didn't take long to overcome, and within hours the students at UCLA were executing programs on the SRI machine. Admittedly, as far as the SRI machine was concerned, initially at least, the ARPANET connection was just another terminal – but 1 October 1969 saw Crocker and Cerf able to report success in the very first connection on what would become the internet. By the next summer, nine machines were live on the network. In 1971, the first network email would be sent. The world may not have realised it (I was still sending punched cards in the mail at this point in time), but things had changed forever.

The wilderness years

Although ARPANET was initially targeted at universities, it was developed with the potential for military applications in mind. In 1983, part of the network was separated off for purely military use, renamed MILNET, while the remainder remained the ARPANET, but became the starting point of the internet as we now know it. The network's growth was dramatic by the standards of the time. By 1988, around 60,000 computers were connected to ARPANET. It was this year that computer operators would get their first experience of a new phenomenon that is now all too familiar, but that would have disastrous results because of the distributed network.

Computers on ARPANET started to slow down for no obvious reason. This phenomenon was spreading almost like a disease, as machine after machine started misbehaving. Operators restarted computers and cleaned up their code, but soon after reconnecting them, they would fail again. Eventually, the whole ARPANET had to be shut down. (It's interesting to think what the implications would be of an equivalent issue that require shutting down the whole internet.)

The problem turned out to be the responsibility of a Cornell University graduate student called Robert Morris. Everyone knew that the ARPANET was big, but it had got to the stage that no one was sure exactly how many computers were connected. Morris devised a program to take a census of connected computers. It was intended to work by using a small fault in the mail program on the predominantly Unix-based university computers. Morris had written what was supposed to be a small, unnoticeable program that passed from computer to computer using email, tallying up a count.

The program checked to see if it was already present before installing itself, but Morris realised that operators might spot it and put a dummy program in its place. So, one in seven installations of his program ran itself even if it was already present. The result, as copies pinged from computer to computer and back again across the distributed network, was hundreds of copies of the program running on each computer, grinding it to a halt. Morris had accidentally written the first computer worm. With a particular irony, a computer operator rang up the National Security Agency and was passed to a person called Robert Morris to deal with the problem – the agent contacted was the father of the student who would later be the first person to be convicted under the Computer Fraud and Misuse Act.

Although the internet made steady progress in academia, and its protocols were adopted by a good number of businesses, this was no sudden, overnight success. Consider, for example, the smartphone, which took off with the launch of Apple's iPhone. Within a handful of years this was a ubiquitous piece of technology. Yet if we fast forward from 1969 to, say, 1995 – 26 years later – the internet hardly existed as far as the majority of domestic users were concerned.

If you wanted to connect up to other computers, you dialled into a private, proprietary network. If you were technically minded, this tended to be CompuServe, which offered a considerable amount of flexibility. For those looking for a more packaged connection, there was AOL (originally America Online Limited), while Apple users had their own ring-fenced eWorld network.

The reason I specifically chose 1995 as a date when the internet was still relatively unknown was that this was the year that Microsoft launched its big step forward in the Windows operating system, Windows 95. At this launch, Microsoft focused entirely on its own new proprietary network called MSN. The internet was dismissed as an academic irrelevancy.

The networks of the day provided email, discussion forums, an early version of online shopping and more. But they were very much constrained by what the individual companies offered. The difference between this and what the internet would bring, thanks to the addition of the World Wide Web to give it a usable framework, was a bit like the difference between watching a single old-fashioned TV network and having the whole modern range of streaming options.

Initially, the web seemed a niche novelty, because it had the mindset of the early, academic internet. Outside its original professional use, it provided the facility to 'visit' a website, usually at an institution, mostly text-based (where there were images, they were crude and very slow to download), and without any great purpose in mind. For all the complaints that there have been from some early internet fans about the commercialisation of the internet as a result of the web, it was that commercial involvement that started giving the internet real value to the general public.

The basic, underlying internet features, such as email, were still there. But with the web it became possible to shop online, to access information in new and different ways. And though some of the big modern uses of the internet such as video streaming and video calling don't themselves necessarily use the web per se, we still get into them via a web interface, before specialist apps making use of internet protocols take over. Techies get irritated when the public confuse the internet and the web, saying, for example, that Tim Berners-Lee invented the internet. He didn't. But there is no doubt that Berners-Lee's World Wide Web was the public face of the internet that made it the success that it has been.

What was once a niche technical communications network has replaced everything from the phone network to paper documents as the universal communications medium. Combined with the incredible processing power of a whole range of devices from mobile phones, through computers, smart TVs, smart speakers and more, this has become far more than a way of exchanging data, transforming the way we run our lives both domestically and for business.

Life changers

The web

The degree to which we now rely on the internet through the World Wide Web could not have been predicted by anyone. From online shopping to satellite navigation, the web has transformed many lives and jobs.

Email

Although email was accessible through other networks before the internet became ubiquitous, it is the standardisation provided by internet protocols that has enabled email – and related applications, such as messaging – to become as universal as the postal system, with near-instant response rates. Both businesses and individuals rely on these systems for many of their communications.

VOIP and video links

The lack of a charging mechanism for the internet initially had limited impact on telecommunications companies. But now, free internet calls using VOIP – voice over internet protocol – are commonplace. This has had a significant impact on particularly international calls, which remain expensive by conventional means, where conventional calls within a country are often now provided as part of an access package. Video calls were significantly slower to take off, despite being long predicted by science fiction, as they can feel uncomfortably artificial. However, the 2020 COVID-19 pandemic saw a blossoming in the use of video calling, particularly to support remote working, and it seems likely that it will now continue at a higher rate of take-up.

Transformation of TV

Because the sheer bandwidth required would have been unthinkable in the early days of the internet, one of the less predicted results of internet proliferation has been the move of television from channels that broadcast on a schedule to on-demand video that can be started as and when a viewer decides. As of 2021 we are at a transitional stage where scheduled channels still have considerable sway. However, it seems inconceivable that such broadcasts will continue for more than a generation, after which it is likely that all television will be streamed.

The cloud

Because we don't see it – and because it is only a virtual entity enabled by the internet – we tend to forget the importance of the cloud. In effect, this is a combination of access via the internet and large-scale data storage facilities. Without the cloud, streaming TV would not be feasible. But also, as we store more and more in digital form (our collections of photographs, for example), the cloud has become a huge safety net. When I first worked with PCs, I was using early devices which were not as reliable as current machines. In the first six months of using IBM's then ground-breaking PC AT, I had two hard disks fail, losing all the data on them. This was an object lesson in making backups, and I have been obsessive about doing so ever since. I even used to store a copy of backup disks at a second location, in case of a fire in the office. But now, anyone can have their data automatically backed up to the cloud at a trivial cost.

Internet of things

The internet was conceived as a way of connecting a relatively small number of large computers. However, we now see a world in which information technology is present in far more devices than those explicitly identified as computers. The most obvious case is the mobile phone, which in smartphone form has become nothing less than a pocket computer. In 2020 there were over 3.5 billion smartphone users worldwide. However, there are also many more devices now being connected to the internet. On a quick survey of my own home, we have at least fifteen internet-linked devices. As well as computers and phones, these include a TV, a printer, central heating and lighting. We are also seeing doorbells, alarms and more joining this 'internet of things'. While some connected devices (toasters and coffee makers, for example) remain more novelty items than of any great significance, this is a move that is not going to go away.

The eleventh day?

It has been a long time since 1969. Things have certainly moved on. When Crocker, Cerf and their colleagues were involved in setting up the second node on the internet, they could hardly have envisaged our modern, hyper-connected world. Yet the basics were all there. There have been many developments in physics since, notably in our understanding of sub-atomic particles, from the acceptance of the sub-structure of protons and neutrons in quarks and gluons to the detection of the Higgs boson. Equally, there have been many advances in cosmology, with the triumph of the big bang theory and the discovery of black holes, dark matter and dark energy. Yet none of these has had significant impact on our everyday lives.

Some have argued persuasively that modern physics has become too dependent on mathematical modelling and not enough focused on reality. The movement started by James Clerk Maxwell, described in Day 4, from mechanical models to purely mathematical ones was not intended to detach physics from reality. And yet, a lot of effort in the modern physics world is arguably wasted on ventures where mathematical 'beauty' is given more emphasis than any tie to observable reality.

A vast amount of effort, for example, has been put into string theory, without any way of demonstrating its validity yet being offered. Concepts such as supersymmetry, which predicts the existence of a whole new raft of particles that have never been observed, still drive proposals to spend more and more on larger particle accelerators, when existing accelerators, which were expected to detect some of these particles, have failed. It's not that physics shouldn't be doing experiments, but rather the directions it is going in – and the expenditure required to do so – seems to have become hidebound in some fields. There has been so much effort expended that physicists are reluctant to give up pursuing what could be a lost cause. Some philosophers of science suggest we may have to wait for a generation of physicists to die out before new ideas can truly emerge.

However, there are still plenty of ways physics and physics-based engineering could have transforming impact on our lives. I'd like to suggest four possibilities – not all ones that I'm convinced will do anything anytime soon, but all with prospects for future transformative capability.

Folding fanatics

The first area is artificial intelligence. This aspect of computer science is already showing impressive potential, whether it is in the ability to help understand protein folding (more on that in a moment), or in providing the necessary support to enable self-driving cars to be relatively safe. Artificial intelligence has been around as a concept since the 1960s and has had many false starts, but it does seem at last that some AI algorithms are making real progress.

One word of caution: AI has always suffered from over-hyping, and this is still the case today. It is true that AI software has had impressive success at playing certain games – but each version of the software is very limited in its application. The protein-folding breakthrough, demonstrated by DeepMind's program AlphaFold in November 2020, was widely trumpeted in the media with headlines such as 'AI cracks 50-year-old problem of protein folding'. There is no doubt that the AlphaFold program has done far better than previous competitors – but the problem is by no means 'cracked'.

Proteins are very large, complex molecules which naturally fold up into shapes that influence their behaviour. Knowing how a protein folds is essential in understanding its function. There are millions of proteins, but the structures of relatively few are known – and it takes a long time to deduce them experimentally. As a result, for a good number of years a competition has been held to see which computer program is best at predicting the structure of a protein. The 2020 breakthrough was that the latest version of AlphaFold beat its competitors handsomely, leaving the rest so far behind that they are hardly worth using. And that is impressive.

However, dramatic though the step forward was, only two thirds of AlphaFold's predictions matched the real structures – and without knowing those structures, we would not know which two thirds were correct. This is not good enough to base, say, a new vaccine on. Even those that were 'correct' were too far away from predicting exact atomic positions within the protein to be directly usable for drug development. That's not to say the program is useless. Its predictions can certainly speed up experimental investigations – but it does not do away with them.

Similarly, AI enthusiasts tell us to expect self-driving cars to be common on the roads any day now. This development has the potential to transform transport in the same way that electronics or the internet have been transformative. Self-driving cars would reduce road accidents – which currently kill over a million people worldwide each year. They have the potential to do away with the need for most of us to even own cars, if one could arrive at the door in a couple of minutes of being requested. And they can cram more traffic into crowded roads by driving closer together, acting almost like trains where the carriages can split off at every junction.

This all sounds wonderful, but the proponents of self-driving cars tend to avoid considering their potential pitfalls. It's true they would significantly reduce road deaths, but these are far more frequent in many regions which are unlikely to be early adopters. For example, if we compare road deaths per million inhabitants in Europe in 2018, the UK was safest with 28, followed by Denmark (30), Ireland (31), Netherlands (31) and Sweden (32). Least safe were Poland (76), Croatia (77), Latvia (78), Bulgaria (88) and Romania (96). Admittedly, the US was significantly worse at 124 per million. But all the top rates were in Africa, the three worst being Central Africa Republic (336), Democratic Republic of the Congo (337) and Liberia (359). Things are even worse if we compare fatalities per million cars, where Somalia has over a thousand times the death rate of the UK.

What's more, although self-driving cars will reduce road deaths, they will still kill people – they already have done so with only a small number on the road. The taxi-replacement company Uber announced at the end of 2020 that they were selling off their self-driving car arm, and this seems likely to have been partially due to bad publicity when an Uber self-driving car killed someone (even though it was not the fault of the AI software). Such deaths are likely to cause a growing backlash. The deaths that are prevented are just anonymous statistics, but the deaths caused by self-driving cars are real people, with families who will blame the technology.

Another issue is likely to be the focus of development on areas like California, where roads tend to be wide and well-maintained, with cities constructed in a logical grid pattern. There is a much bigger challenge, for instance, in dealing with the far older, narrow, twisty roads of Europe, let alone roads in much of Africa and Asia. And then there's the matter of sabotage. It has been demonstrated that a small sticker, hardly visible to a human, applied to a stop sign can fool a self-driving car into thinking it is a totally different sign, letting the car drive straight into a dangerous junction. I don't doubt that self-driving cars will come – but it would not be surprising if it takes until 2050, say, before they are common on our roads.

Artificial intelligence shows no signs of getting anywhere near close to the typical science fiction portrayal of artificial general intelligence – an AI, such as a robot, that can think as flexibly as a human. All the successful applications have been very specific, but that doesn't mean that AI will not continue to grow as an influence on our lives. The same is probably true of our next big thing for physics and physics-based engineering: the transformation of displays.

Living in Glass houses

The way that we see information presented on a screen has changed remarkably since computers became commonplace. In the 1980s, a photograph displayed on a computer screen would probably only have 256 colours and would be at best 640 pixels wide. Now they can use over 16 million colours and are crystal clear. TV screens have become far bigger and sharper, not to mention losing all the bulk of an old TV set. We carry stunning colour screens in our pockets, or on watches. Yet some of the predictions of science fiction have yet to become a reality.

Ever since the 1930s, SF stories have promised us that the future would include television that worked in three dimensions. Yet despite many attempts to provide this, 3D TV still shows no signs of becoming mainstream. Arguably there are two reasons for this: most 3D approaches require the viewer to wear special glasses, and there is limited benefit in terms of the relatively limited immersion we get in the home, when compared with a cinema.

Many of us are prepared to put up with 3D glasses occasionally to watch a film on a big screen (though many still go to the 2D version), but it seems one imposition too far in something as casual as TV watching. It's also true that, if anything, we tend to have a less immersive relationship with our screens than used to be the case. Just as many of us no longer bother with a full-scale stereo system, preferring the convenience of listening on smaller devices, so much viewing is either pushed to a phone screen, or performed while simultaneously doing something else on a second screen. Any value that might be had from 3D TV does not really tie into such a casual mode of viewing.

However, according to the evangelists for transformed display technology, the future is not TV at all, but rather AR/VR – augmented reality and virtual reality, which is only really effective when combined with a headset. Augmented reality involves superimposing virtual video constructs on a view of the real world – perhaps the best-known example is the Pokémon GO game. Virtual reality replaces the entire viewpoint with a computer graphics scene – first-person computer games are the closest many have to experiencing this.

The way most of us have experienced these phenomena is through a phone screen or a games console. But to get the full experience, current technology requires a headset that entirely covers the eyes, which is both clumsy and hard to wear for any length of time. The big breakthrough here is the expectation of being able to get the full AR/VR experience into something as light and unobtrusive as an ordinary pair of glasses from the opticians.

There is no doubt that these technologies are getting better, but like the self-driving cars, there are obstacles that have not been given enough consideration. An early example of AR glasses was Google's Glass project – a pair of glasses that had a camera and projected an image in a tiny, magnified virtual screen onto the view through the lens. There is little doubt that Glass was a flop. It was expensive and looked clunky – but more importantly, there was a significant backlash from others when individuals wore them. Wearers either were mocked or were banned from locations because of concerns about privacy.

The reality seems to be that most of us are reluctant to wear too much technology on our face, and that users of AR/VR glasses may suffer abuse from those around them. Some experts in the field believe that such glasses will be commonplace by 2025, but that seems overly optimistic. Again, like self-driving cars, the technology will continue to advance and will be eventually adopted. We may even see a point where this kind of technology can be crammed into a contact lens. However, it feels that we are probably looking at the 2030s for widespread acceptance of AR/VR glasses.

Computing with particles

Perhaps a closer bet for a limited widespread adoption is quantum computing. Here, the physics goes one step further from electronics in making more explicit use of the strange behaviour of quantum particles. All semiconductor electronic devices depend on quantum principles, but the logic of the computing operates on the normal level of bits, which can have a value of 0 or 1. A quantum computer replaces these with qubits, represented by the states of quantum particles, which can effectively hold more than one value at a time and can operate together in a way that multiplies up the power of the computer.

Labs around the world have been attempting to construct quantum computers for several decades, but the technical challenges are huge, and the physics itself is pushing the edge of our knowledge. Things are changing, though. Experimental quantum computers are starting to get to the stage where they are able to perform a few tasks that would be impossible for existing conventional computers to manage in the same time (so-called supremacy). And there are some algorithms for quantum computers which, with the right level of technology, would enable them, for example, to hugely speed up searching, a powerful limit on current computing technology.

We are not going to see quantum computers on the desktop. This is partly because they are not general-purpose devices like a PC. They are potentially unbeatable at some tasks, but very limited in the range of tasks they can work on. But it is also true that, for the moment, even the very limited quantum computers that exist in laboratories require extreme conditions. For example, many need to be supercooled to near absolute zero – not exactly practical in the home.

Now, though, the cloud provides us a mechanism to have the best of both worlds. Most computers have separate processors for handling graphics. The main processor hands graphic processing over to this specialist unit and gets back the results. The same can be done with a quantum computer in the cloud – an application running on a conventional computer can hand over specialist requirements to a quantum computing unit and get the results back. We can expect to see quantum computers starting to have a significant impact by the end of the 2020s. Certainly well before the final suggestion for the eleventh day.

Electricity that is too cheap to meter

In 1954, Lewis Strauss, at the time chairman of the US Atomic Energy Commission, told an audience, 'It is not too much to expect that our children will enjoy in their homes electrical energy too cheap to meter.' He was not saying that it would be free, but rather could be provided on the same unmetered basis as water (which, ironically, now often is metered). The reason for this optimism was atomic power. But it never seemed likely that this would be the case with the nuclear reactors of the day. Some have speculated that Strauss was, in fact, referring to nuclear fusion energy.

Nuclear fusion is the power source of the Sun. Unlike the current fission reactors, it does not require fuel such as uranium, making use of far less hazardous isotopes of hydrogen as fuel – and it requires far less fuel to generate the same amount of energy. However, fusion is very hard to start and keep working. One of its positives, compared with fission, is that a fusion reactor cannot run away out of control – it will stop of its own accord at the least provocation.

Back in the 1950s they had no idea how difficult it would be to get fusion operating, and then to get it to a state where it gave more energy out than was being put into in the first place. Ever since it was first conceived, fusion has been predicted to be about fifty years from becoming mainstream and it still is. Slow though this may sound, we have made huge progress.

The main hope now is a project called ITER (International Thermonuclear Experimental Reactor), based at Saint-Paul-lès-Durance in Provence, France. Construction started in 2013 with the device expected to be fully fired up by 2025 and the hope that this will be the first fusion reactor to produce more energy than it takes to run. It is still, however, an experimental device and the next generation would be the first that could be sensibly considered as a working power plant, perhaps by 2050, with mainstream adoption another twenty or so years in the future.

Although wind, tide and solar can provide a significant part of electricity requirement, we will always need to even out these variable sources. One possibility – and another transformative physics-based technology in its own right – would be advances in battery technology that make storage far more effective. The other likely backup generation means is nuclear, and it is only likely to be with the development of fusion generators that this can have a long-term future.

Prediction, as Niels Bohr (among a number of others) is said to have remarked, is difficult, especially about the future. The chances are that most of what I've written in this chapter won't be accurate – and it is entirely possible that totally new physics and physics-based technology will come along to make major differences to the way we live. What is certain, though, is that there will be another day to come when the application of physics once more changes lives.

Further Reading

Day 1

Approachable biography of Newton – *Isaac Newton: The Last Sorcerer*, Michael White (Fourth Estate, 1998)

In-depth biography of Newton – *Never at Rest*, Richard Westfall (CUP, 1983)

Gravity – *Gravity*, Brian Clegg (St Martin's Press, 2012)

The *Principia* – *Magnificent Principia*, Colin Pask (Prometheus Books, 2019)

Day 2

Biography of Faraday – *Michael Faraday: A Very Short Introduction*, Frank James (OUP, 2010)

Context of Faraday's electrical work – *Michael Faraday and the Electrical Century*, Iwan Rhys Morus (Icon, 2004)

Day 3

Laws of thermodynamics – *The Laws of Thermodynamics: A Very Short Introduction*, Peter Atkins (OUP, 2010)

Day 4

Biography of Maxwell – *Professor Maxwell's Duplicitous Demon*, Brian Clegg (Icon, 2019)

Day 5

Biography of Curie – *The Curies*, Denis Brian (Wiley, 2005)

Radium mania – *Half Lives*, Lucy Jane Santos (Icon, 2020)

Day 6

Biography of Einstein – *Einstein: His Life and Universe*, Walter Isaacson (Simon & Schuster, 2017)

Relativity – *The Reality Frame*, Brian Clegg (Icon, 2017)

Day 7

Superconductivity: *Superconductivity: A Very Short Introduction*, Stephen Blundell (OUP, 2009)

Quantum applications: *The Quantum Age*, Brian Clegg (Icon, 2014)

Day 8

History of the transistor: *Crystal Fire*, Michael Riordan and Lillian Hoddeson (Norton, 1997)

Day 9

See above (Quantum applications)

Day 10

History of the internet: *Where Wizards Stay Up Late*, Katie Hafner and Matthew Lyon (Touchstone, 1996)

Internet in society: *Tubes*, Andrew Blum (Ecco, 2012)

Picture credits

THE LIFE AND SCIENCE OF
JAMES CLERK MAXWELL

PROFESSOR

MAXWELL'S

DUPLICITOUS

DEMON

BRIAN CLEGG

James Clerk Maxwell transformed the way physics was undertaken. His explanation of the interaction of electricity and magnetism revealed the nature of light and laid the groundwork for everything from Einstein's special relativity to modern electronics.

Along the way, he set up one of the most enduring challenges in physics. 'Maxwell's demon' is a tiny but thoroughly disruptive thought experiment that suggests the second law of thermodynamics, the law that governs the flow of time itself, can be broken.

This is the story of a groundbreaking scientist and his duplicitous demon.

ISBN 9781785785702 (paperback) / 9781785784965 (ebook)

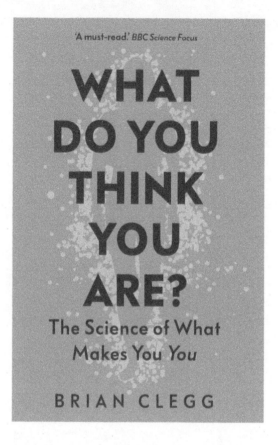

'A must-read.' *BBC Science Focus*

WHAT DO YOU THINK YOU ARE?

The Science of What Makes You You

BRIAN CLEGG

What makes you the unique individual that you are?

From the atomic level, through life itself to consciousness, genetics and personality, *What Do You Think You Are?* explores how each aspect of you – your DNA, your memories, your flesh and bone – has come to be.

Full of fascinating true stories – featuring royal ancestors, stellar deaths, real-life hobbits and a self-reproducing crayfish, to name a few – this wide-ranging exploration of what makes you *you* is a one-of-a-kind voyage of (self) discovery.

ISBN 9781785786600 (paperback) / 9781785786242 (ebook)